B45.00

The Fruit, the Tree, and the Serpent

THE FRUIT, THE TREE, AND THE SERPENT

Why We See So Well

LYNNE A. ISBELL

Harvard University Press
Cambridge, Massachusetts, and London, England   2009

Library of Congress Cataloging-in-Publication Data

Isbell, Lynne A.
The fruit, the tree, and the serpent : why we see so well /
Lynne A. Isbell.
p.   cm.
Includes bibliographical references and index.
ISBN 978-0-674-03301-6 (alk. paper)
1. Human evolution.   2. Primates—Evolution.   3. Vision.
4. Eye—Evolution.   5. Snakes.   6. Fear.   I. Title.
GN281.4.I82   2009
599.93'8—dc22   2008037456

May we have the vision to last another 80 million years

Contents

Preface

If you are holding this book in your hands and you are reading these words, you can thank your primate ancestors. Of all mammals, only primates can both grasp objects in their hands and see so well that they can distinguish easily between the individual squiggles that we humans use to create our written words. If you have gone as far as this sentence, you are probably also fairly curious about the world around you. Again, you can thank your primate ancestors; the evolution of our large brains, which give us our curiosity, really took off with them.

Part of our curiosity about the world is personal; who is not interested in their relatives, if for no other reason than that they tell us more about who we are? For some of us, this extends only as far as our closest relatives, or perhaps to those we did not know but can learn about by tracing our family trees as far back as several hundred years. But there is a greater intellectual curiosity about who we are that, once whetted, cannot be satisfied by our own individual histories. We begin to pursue who our relatives were much farther back in time, even before the dawn of our species. And then we need to know why they were *our* relatives and not someone else's. Professors of biological anthropology commonly engage in this sort of intellectual pursuit, and I am no different. But for much of my career my curiosity about what happened in deep time did not extend beyond the point of about six million years ago when our primate ancestors first began to walk bipedally, moving around on two legs instead of four. Ah, to be a hominin! What special creatures we were to walk upright! I am a primate behavioral ecologist trained in animal behavior. I focus on primates for their fascination in the present day. But because I work in an anthropology department, I thought it especially important to know about the evolution of our immediate ancestors. Yet my own time constraints and the volume of information I needed to digest about the behavior and ecology of primates existing today made me think I could not afford to learn details about our relatives that lived before six million years ago.

That changed by fortuitous accident one day in 1999. At the time, I was pursuing the answer to why female primates living in Central and South America (the New World) are more willing to leave their relatives (social dispersal) and their home areas (locational dispersal) than those living in Africa and Asia (the Old World). (Yes, it is a rather esoteric question, not one that many people wonder about, but curiosity is rampant in academia.) Because locational dispersal is known to put dispersers at greater risk of predation, I was investigating the possibility that primates in the New World might have less predation pressure than primates in the Old World, making it less costly to disperse. The main predators of primates nearly everywhere are cats (the felids), raptors, and snakes. My own field research had shown that felids, particularly leopards *(Panthera pardus)*, can annihilate entire primate groups (much to my dismay), and so I naturally focused first on where and when the cats evolved. I soon discovered that felids evolved in the Old World, and fairly recently as evolutionary time goes. They entered South America even more recently, about three million years ago when the isthmus of Panama formed between North and South America. Because primates found their way to South America (the New World) via Africa about 30–35 million years ago, much earlier than the felids, this boded well for my idea that predation pressure has been a weaker evolutionary threat to female New World primates than to female Old World primates.

In my readings, I was intrigued to find that some cats and some primates share a retrovirus that can be traced to about six million years ago (Wayne et al. 1989). We know from studies of another retrovirus, the human immunodeficiency virus (HIV), that retroviral transmission typically involves contact with blood or other bodily fluids, and so I took this information as indirect evidence that six million years ago a cat of some kind bit into an infected African monkey. Intrigued by a reference to the same phenomenon in snakes, I followed up that lead to learn that the Russell's viper *(Daboia russellii)*, a venomous snake from Asia, has a retrovirus more closely related to a retrovirus found in primates of Asia than in primates of South America (Andersen et al. 1979). I already knew that constricting snakes, the pythons and boas, are predators of primates, but finding out that primates and poisonous snakes might also share a long antagonistic history made me wonder where and when venomous snakes evolved. Could it be that they also arrived late in South America, further reducing New World primates' risk of dying?

Thus began what was to become a ten-year near obsession to solve what gripped me as no murder mystery had ever done before. I set aside my focus on dispersal. I broadened my original question from what makes female New World primates different from female Old World primates to what makes primates on the landmasses of Central and South America, Africa and Asia, and Madagascar so different from each other. I then expanded the question further to what makes primates different from all other mammals. That led to a search for what makes mammals different from other vertebrates. Ultimately, my quest took me back to the origin of vision itself. I had to go back that far to understand primates because their vision is what truly distinguishes them from all other animals. Whereas before I thought I had no time to go outside my specialty of behavioral ecology, now I felt I could not stop until I either hit a wall that told me the idea that snakes were very important in the evolution of primates was wrong or I satisfied myself that there was enough evidence (albeit indirect) to support such an idea. Throughout this process, I learned a lot more about primates than I ever anticipated, and I also learned a lot more about other fields and subjects relevant to their evolution, including geology, phylogeography, paleobiogeography, mammalian evolution, molecular systematics, neuroscience, snake evolution, and vision. I even learned a little bit about the non-scientific topics of myths and religions.

Now I find myself compelled to share what I have learned with readers who are curious enough to venture with me into such seemingly disparate scientific disciplines. They do connect. Importantly, delving into these many disciplines has allowed me to see patterns that would be missed with studies of narrower focus. Like a camouflaged snake in the grass, the evidence is out there but we have to look hard to find it because no one has explicitly tested the theory yet. I have done my best to illuminate the evidence for you, and I hope you will see it, too.

The Fruit, the Tree, and the Serpent

CHAPTER ONE

Introduction

CONSIDER THE Old Testament story of Adam and Eve and the origin of humans. Eve's big mistake was that she noticed the Serpent. If she had not detected it, if she had not *looked* at it, she would not have entered into the conversation that tempted her to pick the fruit off the Tree of Knowledge of Good and Evil, and God would not have punished her. (There seems to be some disagreement over what that fruit was, with some people arguing for apples and others for pomegranates. No matter, although as you will come to understand later on in the book, I think it is an interesting aside that we do seem to favor *red* fruits in this story.) It was Eve's attention to the Serpent that did her in. Alas, God punished Eve for noticing the Serpent and eating that fruit by decreeing, among other punishments, that all women would henceforth have pain during childbirth.

Creation stories are often ways to understand and explain our world. This particular creation story invokes snakes, visual attention to them, trees, and consumption of fruits to explain not only women's pain during childbirth but also our curiosity, our morality, and our fear of snakes. The story of Eve is actually quite similar to what I believe the scientific evidence suggests for the origin of primates, including humans. We know today that pain during childbirth is the result of our extraordinarily bigheaded babies, the price we pay for having brains with a large neocortex. Because areas either directly or indirectly involved in vision occupy approximately half of the neocortex, the behaviors associated with vision and our large brains do indeed cause us to give birth in pain.

One of the most accepted hypotheses for the origin of primates suggests that visually guided reaching and grasping (Eve's plucking the fruit) was responsible for separating primates from other mammals (Cartmill 1972, 1992). I will buck the trend, however, and marshal the evidence that although visually guided reaching and grasping may have helped, it was not directly responsible for the acute vision and large brains of primates. The connection between vision and reaching and grasping in primates developed instead in a subcortical part of the brain to give them highly responsive and automatic hand-eye coordination, a skill that would be especially useful during travel in an arboreal environment (Eve's Tree) or perhaps during feeding. I will argue that a more singular preadaptation for acute vision and large brains was the enjoyment of a diet of fruits and nectar (Eve's fruit). When snakes (the Serpent) appeared, a particularly powerful selective pressure that favored expansion of the visual sense, only ancestral mammals having a diet of fruits or nectar from flowers in that arboreal milieu could afford to expand their visual systems, as such expansion typically occurs at a cost to olfaction. Only those mammals that ate odoriferous foods such as fruits and flowers could afford to reduce their sense of smell and still find their food efficiently.

Judeo-Christian religions are not alone in placing snakes in a prominent position in their creation stories. Australian aboriginal creation stories describe the Rainbow Serpent as a creator of humans. Similarly, a Chinese creation story holds that Nugua, a dragon-like goddess, created humans in her image except that she gave them two legs rather than have them glide along as she did (Rosenberg 1994).

Creation stories aside, throughout the world snakes are often held responsible for hardships that people endure. Snakes as troublemakers are found among cultures as diverse as those in India, North America, and Africa. In the mythologies of many cultures, snakes destroy environmental perfection, sleep with goddesses and mortal women, cause droughts and famine, and make people suffer death, as in the case of Adam and Eve (Leeming 1994; Rosenberg 1994; McNamee 2000). For example, a venomous serpent killed the mighty Thor, god of thunder in Norse mythology (Rosenberg 1994).

The preeminence of snakes as evil, dangerous, or at the very least troublesome continues today in our folklore in the making. Indiana Jones feared snakes over all other animals, and they had a way of appearing at the most inopportune times when it was difficult or impossible for him to immediately escape them. More recently, He Who Must Not Be Named (I am not afraid so I will say it: Lord Voldemort), the good wizarding world's most feared presence and Harry Potter's greatest enemy, morphed early on into a snake and used one as his totem (Rowling 1997). Why do modern writers continue to choose snakes to represent evil when we know intellectually that snakes are just snakes?

The prominence of snakes in religions, myths, and past and present folklore around the world underscores our deep human connection to them. This preoccupation with snakes extends from the spiritual categorizations of the world and human behavior via religion and myth to the presumably rational categorizations of the more recently developed social sciences. In Freudian psychology, for example, snakes are a symbol of the penis, a body part of such great importance that 50% of us are supposedly envious of the other 50% who proudly sport one.

Fear of Snakes

A more practical approach of psychology deals with phobias, fears that can incapacitate people who have them. The most common fear is ophidiophobia, which is, you guessed it, the fear of snakes. Of 1,015 adults surveyed in a Harris poll taken in 1999, 36% considered themselves very afraid of snakes (Taylor 1999). Fear of flying came in at a distant 14%. These two fears were campishly exploited recently in the much-anticipated B-movie *Snakes on a Plane*. Coincidentally, the anticipation and public discussion of this movie hit fever pitch at about the time that the synopsis of my Snake Detection theory was published in the *Journal of Human Evolution* (Isbell 2006a). I think the publicity the paper received from Internet news services was heightened by the buzz surrounding the movie. We had snakes on our brains in the summer of 2006.

By definition phobias are largely irrational, at least in their debilitating severity. Rational fear of an object is expressed only after someone has had a negative experience with that object, either directly or indirectly through the behavior of others, but most of us have not had any experience with snakes, good or bad, because we have eliminated them from our surroundings. Because snakes have never posed a threat to most of us, being so fearful of them does not make sense. Try telling that, however, to people with ophidiophobia: intellectually they might agree but emotionally they cannot. There are plenty of people who have not had one bad experience with snakes, but show them a picture of a snake, or worse, a live snake, and they react with elevated blood pressure, increased heart rate, and sweaty palms.

Why do we give snakes such great importance in our religions, myths, and stories? Why do we think of them out of proportion to their reality in our daily lives? Why do some of us fear them so much? It cannot be because of their present danger. After all, more people in the United States are maimed and killed by cars every day than by snakes over an entire year, yet few people have a car phobia. It cannot even be simply because of the danger they posed to us in our ancestral homeland. I have worked for many years in East Africa, the homeland of our ancient hominin ancestors, and have avoided stepping on most of the snakes I have seen, but I have also had close encounters with lions *(Panthera leo)*, African buffalo *(Syncerus caffer)*, and elephants *(Loxodonta africana)*, all of which could kill me as thoroughly as a puff adder *(Bitis arietans)* or a spitting cobra *(Naja nigricollis)*. I especially fear buffalo because one nearly did kill me! More people are killed in Africa by elephants and hippos *(Hippopotamus amphibius)* than by other mammals, and these big creatures have been a problem for hominins for millions of years. Assuming snakes and all these other animals generated fear in our hominin ancestors, why aren't lion, elephant, hippo, or buffalo phobias now as common among us as snake phobia?

It may be because this singular fear of snakes goes way back, even farther back than six million years when our hominin line first appeared. Like humans, many monkeys react fearfully when they see snakes, even those with no previous negative experience with snakes. Monkeys born and raised in captivity, where the only snakes are harmless, still react strongly when they see one. I recently developed a study that called for exposing captive-born-and-raised rhesus macaques *(Macaca mulatta)* to a model of a snake. The day I began my study, I parked myself at one of the many outdoor enclosures of the California National Primate Research Center and waited with a camera to photograph the monkeys' reaction first to a cloth and then to a model snake uncovered by the cloth. Eight minutes into the beginning of my study, I was alerted by the monkeys' focused attention to something on my left. I was surprised to see a gopher snake *(Pituophis catenifer)* at least five feet long moving slowly toward their enclosure. Although I knew such snakes lived in the area, in all the years I had spent watching monkeys at that primate center, I had never seen a live one there. What luck to have an actual snake approach the very cage I was observing, to be only a few meters away from it, and to have a camera available to photograph the reactions of the monkeys.

Although the monkeys were somewhat interested in the snake model, the real thing drew a much stronger reaction. In an enclosure holding about 80 monkeys, at least 30 gathered around in a mob to watch the snake, some on the ground, others clinging to the fence. As the snake progressed into the enclosure, the mob parted like the Red Sea for Moses, giving it a wide berth to pass through unhindered. Snakes elicit fearful fascination in captive-born-and-raised rhesus macaques who likely have never been harmed by one. This suggests that ophidiophobia may go way, way back, to at least 30–35 million years ago when the first Old World monkeys and apes, the so-called catarrhine primates, are thought to have appeared. Ophidiophobia may even extend farther back to 60 million years ago when the first generalized simian primates, the anthropoids, are thought have appeared. If so, this time line might help explain the shared ophidiophobia of all anthropoids, including humans.

Seen in this light, perhaps it does not seem so odd that, although these same monkeys grow up learning that humans can be dangerous, they do not develop a fear of all humans. They do not react fearfully toward me or other researchers who only observe their behavior. They react fearfully only toward the animal technicians who sometimes need to remove individuals from the social group and thus are the equivalent of predators in the natural world. The monkeys' fear of humans is a rational, selective fear, not a phobia that generalizes to all humans in all situations. We humans have not been dangerous to other primates long enough for that.

Constricting Snakes and Mammalian Evolution

The fear of snakes likely has an even longer evolutionary history than 60 million years. I can say this because the molecular evidence suggests that modern placental mammals evolved around 100 million years ago (Springer et al. 2004), and, as you will see, all modern placental mammals today share a common set of brain structures that help them avoid objects, such as predators, that are dangerous to their survival. This set of structures may help to form the neurological basis for what psychologists Arne Öhman and Susan Mineka have called the "mammalian fear module" (Öhman and Mineka 2003). The fear module was likely present in the earliest mammals in response to their first predators; those that did not respond to predators by detecting and executing appropriate motor behaviors to avoid them clearly would not have survived for long. We know that the main predators of modern mammals are snakes, birds of prey, and carnivorous mammals such as wild relatives of our cats and dogs. I will show you that among such predators, boas and pythons, constricting snakes, were the first to evolve and that they likely evolved in part in response to the presence of the new mammals, which were a novel food source.

There are 18 orders of modern placental mammals of which primates are one. In the past, the relationships among those orders were defined by their similarities in physical characteristics. Partly because of similarities in their visual systems, the closest relatives of primates have long been considered to be bats, treeshrews, and colugos (also called flying lemurs, a misnomer because they do not fly and they are not lemurs). Together, these animals were referred to as Archonta. Recently, however, these higher-level relationships have been reassessed as a result of tremendous growth in molecular systematics. In common with morphological studies, molecular findings consistently place primates into a taxon with treeshrews and colugos but they differ from morphological studies in that they exclude bats from this taxon. Surprisingly, bats are more closely related to whales and bears than to primates. There have been other twists as well. Molecular studies also reveal that rats and rabbits are more closely related to primates than morphological studies suggested. Together, the group of orders uniting primates, colugos, treeshrews, rabbits, and rats is now referred to as Euarchontoglires. I will employ these new molecular findings to challenge current views on the time and place of the origin of primates and to propose an alternative.

Although the fear of snakes likely goes farther back in the mammalian lineage than primates, I will provide evidence that primates have uniquely enlarged the fear module and incorporated vision more extensively into it. If we are to understand why we primates uniquely evolved the combination of excellent vision and large brains in response to snakes, however, then we also have to figure out why our closest relatives, the colugos, treeshrews, rabbits, and rodents, indeed, all other mammals, did not also evolve those traits. The question of why snakes did not similarly affect other mammals is essential to answer if the theory that snakes were responsible for primate origins is to have any credibility.

Venomous Snakes and Variation in Primate Vision

Until now I have been describing primates as a uniform group of mammals. Even within primates, however, there is tremendous variation in vision and brain size. Compared with other primates, Malagasy prosimians (the lemurs) have relatively good olfaction, small brains, and poor vision, even those that are active during the day. In contrast, catarrhine primates, the apes, humans, and Old World monkeys, have poor olfaction, large brains, and excellent vision (including the unique ability among mammals for both males and females to distinguish between red and green). Is it just a coincidence that poorer vision and smaller brains are found in those primates that have never coexisted with venomous snakes and that excellent vision and large brains are found in those primates that have always coexisted with venomous snakes? Perhaps, but the coincidence becomes more suspicious when we factor in that platyrrhine primates, the New World monkeys, have had more exposure to venomous snakes than Malagasy prosimians but less exposure than catarrhine primates, and that they also have visual systems that are intermediate between those of prosimians and catarrhines. Is this still merely a coincidence, or does the variation that exists today in the brains and vision of primates at broad taxonomic and geographic levels actually have an evolutionary explanation, e.g., could the variation be a direct result of variable coexistence with venomous snakes?

I will provide evidence that while constricting snakes began the evolutionary arms race with their mammalian prey, venomous snakes continued it and shaped it in extraordinary ways. As is typical of evolutionary arms races, the prey then responded. Mammals that did not have the preadaptation of feeding on high-sugar foods in arboreal habitats responded with various kinds of physiological resistance to venoms. In contrast, mammals that had invaded that ecological niche early on were able to respond with gradual improvement of their visual systems until what we have today are primates with an incredible visual ability by mammalian standards. Not only are we catarrhine primates able to detect snakes barely visible in the vegetation but we are also really good at choosing the ripest, sweetest fruits by color alone. Some of us are even able to create gloriously colorful paintings, master the arts of fine needlepoint or microsurgery, or read the fine print on legal contracts.

I hope that by the time you get to the final chapter you will be convinced of the possibility that snakes could have been largely responsible for the establishment of primates through their effects on primate vision, and for the differences in vision within primates. Admittedly, it is a long jump in time from the beginnings of catarrhine primates to our own hominin lineage but the extreme fear of snakes persists in us and presumably was also present in our hominin ancestors. Could snakes have affected hominins in any other way? In the epilogue, I speculate that they could have influenced the development of a seldom recognized but distinctively human behavior: our ability to point to objects, say, an airplane in the sky, for the purpose of drawing others' attention to those objects. Lacking any alternative hypothesis, perhaps this idea will generate more systematic attention in the future.

As a result of my research on snakes and primate vision, I have come to see that the heady position of snakes as our destroyers and creators in religion, myth, and folklore has not been too far off from what the scientific evidence shows. The goal of this book is to provide you with scientific evidence that snakes have indeed been important destroyers and creators in primate lives and consequently in our own evolution. It will offer you a new way of looking at our primate origins that I think makes good sense of who we really are, revealing the ultimate reasons behind both our deepest fears and our strongest, most persistent, and most extraordinary attributes.

CHAPTER TWO

Primate Biogeography

Our greatest tool is, of course, our ability to extrapolate from the consistent patterns we see among living species to these more poorly known animals in the fossil record. We must keep an open mind, however; the fossil record is likely to be full of unique events.

John Fleagle, *Primate Adaptation and Evolution* (1999: 316)

To UNDERSTAND the origin and past distribution of primates, it is necessary to understand what modern primates are like and where they live today. What makes a primate different from a rodent, rabbit, or treeshrew, and do they live today where they lived in the past?

Modern Primates

All primates share a suite of characteristics: grasping hands and feet, nails on at least the first toe, eyes that are more or less directed to the front of the face, reliance on vision as the predominant sense, and enlarged brains (Dagosto 1988; Martin 1990; Cartmill 1992; Rasmussen 2002a). Some of these traits are shared with other mammals but no other mammals have them all. Rodents share perhaps the greatest number of traits with primates. Among the rodents, for instance, squirrels are diurnal, rely heavily on vision, and have fairly well developed visual systems (Van Hooser and Nelson 2006). Arboreal rodents that move along small-diameter branches and lianas commonly have a nail on an opposable first toe (Walker et al. 1975). Rodents can also grasp and manipulate objects, including arthropods, using the tips of their digits (Whishaw et al. 1998; Iwaniuk and Whishaw 2000; Whishaw 2003). Rodents also overlap with primates in their hand proportions (the size of some of the finger bones relative to size of the bones in the palm), and easily grasp objects (Hamrick 2001), whereas colugos and treeshrews, mammals that share a more recent common ancestor with primates, do not have the same hand proportions nor the same highly developed grasping ability. The ability to perform skilled forelimb movements, defined as the ability to reach for objects with a forelimb, hold them in a hand or forepaw, and manipulate them with the digits (Iwaniuk and Whishaw 2000), may have developed separately in rodents and primates, or it may have been an early trait of Euarchontoglires, remaining in rodents and primates while declining in rabbits, colugos, and treeshrews after they diverged from their common ancestors with rodents (in the case of rabbits) and primates (in the case of colugos and treeshrews).

In fact, the motor patterns of rodents' forelimbs are so similar to those of primates that rodents are used as models to examine problems in reaching and grasping in humans with Parkinson's disease (Cenci et al. 2002; Whishaw et al. 2002), an affliction with no known cause that results in loss of control over one's movements. Interestingly, however, rodents are not good models for examining *visual control and guidance* of reaching and grasping. Rodents use their auditory, olfactory, and tactile senses to locate and reach for food but they do not need to use their visual sense. Rats without vision are able to locate and reach for food as quickly as they did before becoming blind, but rats without a sense of smell are slower (Whishaw 2003). In studies designed to better understand Parkinson's disease, rats experimentally given symptoms of the disease were unable to correct errors in reaching and grasping even though they could see what they were doing. This is not the case with the primates that have been tested. Humans and monkeys with Parkinson's disease are, in fact, able to correct errors visually while reaching for small items. Although the afflicted monkeys may start out with the wrong reaching trajectory, they are able to modify their movements to avoid hitting obstacles, and afflicted humans are able to shape the hand to grasp items correctly when they can see it close to the items (Pessiglione et al. 2003; Schettino et al. 2003, 2006; Vergara-Aragon et al. 2003; Whishaw 2003).

The one characteristic that rodents and other mammals do not share with primates is heavy reliance on vision. The expansion and specialization of this sense has resulted in the other changes that immediately identify animals as primates, even for novices in natural history. A mammal is readily recognizable as a primate if it has a relatively small snout, eyes that face forward, and a fairly large head that encloses a fairly large brain. The primate brain has become larger in part to accommodate the expansion of the visual system, even to the extent of influencing parts of the cerebral cortex that are not the main visual areas in the brain. For instance, the dorsolateral prefrontal cortex (DLPFC) and posterior parietal cortex (PPC) are areas in the brain that are involved in skilled hand movements.

Given what you have just read about the importance of vision in providing primates with feedback to make skilled hand movements, it should not be surprising to find out that the DLPFC and PPC have extensive connections to visual areas of the brain. As it is generally true that the brain devotes more volume to functions that are more important to the owner's lifestyle and survival, it is not surprising that the DLPFC and PPC are larger in primates than in other mammals (Preuss 2007).

The primate groupings of interest in this book are three large taxa: prosimians, platyrrhine monkeys (New World monkeys), and catarrhine primates (Old World monkeys, apes, and humans) (Figure 2.1). These are natural groupings because although all primates share some physical characteristics, each of these taxa has distinctive characteristics that evolved as a result of isolation on different landmasses. Platyrrhines and catarrhines are collectively called anthropoid primates (Table 2.1).

Madagascar is home to most prosimians; relatively few others live in Africa and Asia. Of all the primates, prosimians look the most like other mammals. First, they have whiskers and generally larger snouts than anthropoids. This latter difference reflects greater reliance on olfaction, which prosimians use for the same purposes as other mammals: identifying territories, inspecting the markings of others, determining if females

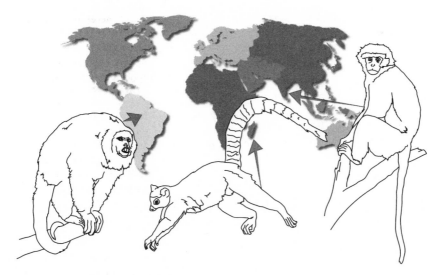

Figure 2.1   Representative primates and where they live. Platyrrhines are found in the New World (South and Central America) (left). Prosimians are found mainly in Madagascar, but there are prosimians also in Africa and Asia (middle). Catarrhines are found in Asia and Africa (right).

Table 2.1 Classification of the primates (all living primate genera included in the text are listed).

Order	Suborder	Infraorder	Common Name
	Prosimii (prosimians)	Lemuriformes (lemurs)	Ring-tailed lemur *(Lemur)* Brown lemur *(Eulemur)* Ruffed lemur *(Varecia)* Mouse lemur *(Microcebus)* Woolly lemur *(Avahi)* Bamboo lemur *(Hapalemur)* Sifaka *(Propithecus)* Aye-aye *(Daubentonia)*
		Lorisiformes (lorises and galagos)	Loris *(Loris)* Slow loris *(Nycticebus)* Potto *(Perodicticus)* Galago *(Galago, Otolemur)* Angwantibo *(Arctocebus)*
Primates		Tarsiiformes	Tarsier *(Tarsius)*
	Anthropoidea (anthropoids)	Platyrrhini (New World monkeys)	Marmoset *(Callithrix)* Tamarin *(Saguinus)* Goeldi's monkey *(Callimico)* Squirrel monkey *(Saimiri)* Capuchin *(Cebus)* Owl monkey *(Aotus)* Titi monkey *(Callicebus)* Saki monkey *(Pithecia)* Howler monkey *(Alouatta)* Spider monkey *(Ateles)* Woolly monkey *(Lagothrix)* Muriqui *(Brachyteles)*
		Catarrhini (Old World monkeys and apes)	Macaque *(Macaca)* Patas monkey *(Erythrocebus)* Vervet *(Cercopithecus)* Talapoin *(Miopithecus)* Baboon *(Papio)* Colobus *(Colobus, Procolobus)* Leaf monkey *(Presbytis)* Gibbon *(Hylobates)*
			Orangutan *(Pongo)* Chimpanzee *(Pan)* Gorilla *(Gorilla)*
			Human *(Homo)*

are in estrus and ready to mate, and finding foods and smelling them for their palatability. Second, most prosimians have prominent ear pinnae, which help them to localize sounds and zero in on their arthropod prey, especially at night. Third, compared with anthropoid monkeys and apes, the bony orbits into which most prosimian eyes fit are situated more to the sides of their faces, although not to the extent shown by most other mammals. These orbits are characterized by a bony bar, which may help to stabilize the eyes or increase visual acuity (Fleagle 1999; Noble et al. 2000; Heesy 2005). Finally, prosimian brains are simpler than those of anthropoid primates, with fewer sulci and gyri, the folds and bulges that are on the surface of primate brains. (A note about tarsiers [*Tarsius* spp.]: they are considered prosimians but they also share some traits with anthropoids, such as a bony orbital plate behind the eye instead of an orbital bar. Thus, in an alternative classification scheme, strepsirrhines are prosimians minus the tarsiers, and haplorhines are anthropoids with tarsiers.)

Like Darwin's finches on the Galápagos Islands, lemurs have evolved into a large number of species and have filled a number of ecological niches, but in contrast to Darwin's finches, they have done all this on one island. Some Malagasy prosimians, such as the woolly lemur *(Avahi laniger),* are nocturnal, others, such as the ring-tailed lemur *(Lemur catta),* are diurnal, and still others, such as the brown lemur *(Eulemur fulvus),* are cathemeral, able to be active whether it is day or night. Some Malagasy prosimians eat a specialized diet of bamboo (bamboo lemurs, *Hapalemur* spp.), some feed heavily on insects and other arthropods (the aye-aye, *Daubentonia madagascariensis*), and others eat mainly leaves, fruits, or nectar (sifakas, *Propithecus* spp., and ruffed lemurs, *Varecia variegata*). Some Malagasy prosimians walk and run across the tops of branches (e.g., brown lemurs), while others leap and cling to vertical trunks and branches (sifakas). A few even spend some time on the ground (sifakas, ring-tailed lemurs). The few non-Malagasy prosimians that live in Africa (the potto, *Perodicticus potto,* and galagos, e.g., *Galago* spp.) and Asia (e.g., the slow loris, *Nycticebus coucang,* and tarsiers) are all nocturnal.

As a group, anthropoid primates (the simians, as opposed to prosimians) differ from prosimians in having smaller snouts (although there are exceptional anthropoids, e.g., baboons [*Papio* spp.], that have very long snouts) and less reliance on smell, more frontally directed orbits for the eyes, greater visual specialization, and larger brains. The postorbital bar of the non-tarsier prosimians is now entirely filled in behind the eye by a bony plate, fully enclosing the orbit behind the eye, and there are no specialized vertical clingers and leapers among the anthropoids. These shared characteristics of platyrrhines and catarrhines likely exist because they once shared a common ancestor. Nonetheless, they also differ in ways that reveal their subsequent geographic separation from one another. Platyrrhines now live in Central and South America and are often referred to as New World monkeys. Catarrhines live in Africa and Asia and are often referred to as Old World monkeys and apes.

The easiest way to tell if a monkey is a platyrrhine is to look at the nose. If it is flat, with the nostrils pointing to the sides, it is a platyrrhine. And if you ever get the chance to look inside a platyrrhine mouth, count the molars and premolars. Starting from the midline at the front of the jaw, after the first three teeth (two incisors and one canine), the next three teeth are premolars and, in most platyrrhines, the final three teeth are molars. (Marmosets [*Callithrix* spp.] and tamarins [*Saguinus* spp.] are alone in having two molars.)

Platyrrhines are more diverse than catarrhines in their behavior and morphology. While most platyrrhines are diurnal, owl monkeys (*Aotus* spp.) are usually nocturnal. In fact, they are the only anthropoids that are active at night. As a whole, platyrrhines have a broad range of diets, but individual species often specialize on gums, fruits, seeds, insects, or leaves. Marmosets cling to trunks of trees as they gouge the bark for gum and have regained claws on most of their digits, which help them to cling better. Some platyrrhines, such as woolly monkeys (*Lagothrix* spp.), have prehensile tails that can hook around branches as a fifth limb to help them feed and move. Others scurry on all fours (quadrupedalism) across branches (e.g., squirrel monkeys [*Saimiri* spp.]). Although not committed vertical clingers and leapers, some frequently leap from trunk to trunk (e.g., Goeldi's monkeys [*Callimico goeldii*] and saki monkeys [*Pithecia* spp.]) (Garber et al. 2005). Still other New World monkeys, such as spider monkeys (*Ateles* spp.), use their arms to suspend themselves below branches while they feed and move about (called suspensory locomotion). Strangely, despite their wide variation in diet, activity cycle, and locomotion, no platyrrhine species habitually walks on the ground.

In contrast to the flat-nosed platyrrhines, catarrhine nostrils point forward or down. As with platyrrhines, catarrhines have three molars on each side of their upper and lower jaw, but only two premolars before the molars. Thus, unlike New World monkeys, which have a total of 12 premolars, top and bottom, Old World monkeys and apes have only eight. We humans fit the catarrhine pattern of nostril position and number of premolars. Representative catarrhines include hamadryas baboons *(Papio hamadryas)*, bonnet macaques *(Macaca radiata)*, patas monkeys *(Erythrocebus patas)*, black-and-white colobus monkeys *(Colobus guereza)*, chimpanzees *(Pan troglodytes)*, and orangutans *(Pongo pygmaeus)*.

As far as behavior is concerned, catarrhines are fairly uniform. All catarrhines are diurnal. Most have a diverse diet of fruits, leaves, and insects, although some leaf monkeys *(Presbytis* spp.) and colobus monkeys specialize in eating seeds or leaves. Most also walk quadrupedally, and while some are habitual tree-dwellers, others are quite at home on the ground. Apes are the only catarrhines that use suspensory locomotion. There are no vertical clingers and leapers or trunk-to-trunk leapers among the catarrhines.

Fossil Primates

Because all primates, even prosimians, have at least a postorbital bar and more or less frontally directed bony orbits, fossilized orbits are one of the characteristics (along with teeth and jaw fragments) used to identify ancient primates. Such fossils show us that at one time primates lived much farther afield than they do now. Fossil primates have been found in ancient rocks in North America, Europe, Asia, Africa, and South America. Interestingly, they have not been found yet in Madagascar, and as I will discuss shortly, this biogeographic pattern has greatly influenced our thinking about the location of origin for primates.

Prosimians were common and widespread in North America and Europe after about 55 million years ago, with fewer in Asia and northern Africa. These fossils reflect the connections of the regions with one another at various times while prosimians flourished (Fleagle 1999). Prosimians diversified considerably during the Eocene epoch (54–34 million years ago); no less than 60 genera have been identified (Fleagle 1999; Covert 2002). Not surprisingly, prosimians were also diverse in their ecological niches. Judging from the size of their orbits, some were apparently nocturnal and some diurnal. Judging from their teeth, they also had a variety of diets, including fruits, leaves, and arthropods. Their postcranial skeletons reveal a range of locomotor types, including clinging and leaping, quadrupedal running, and quadrupedal climbing. Body sizes ranged from very small *(Pseudoloris,* estimated to have been 45 grams) to fairly large (a species of *Notharctus* is estimated to have been about seven kilograms) (Fleagle 1999; Covert 2002). These characteristics are similar to those of modern-day prosimians. Prosimians fade from the fossil record in North America and Europe during the Eocene and Oligocene epochs (34–23 million years ago) (Fleagle 1999).

Inferences based on the fossil evidence suggest that the first anthropoids were tiny creatures that evolved in the Old World, either Africa or, more likely, Asia, at least 60 million years ago (Ross 2000; Beard 2002, 2004; Dagosto 2002; Eizirik et al. 2004; Ross and Kay 2004). Fossils of early, generalized anthropoids have been found in both Asia and Africa but not North America despite the intensity of investigation there (Fleagle 1999). By at least 34 million years ago, anthropoid primates began to differentiate into catarrhines and platyrrhines. Fossils of anthropoids dated to about 34 million years ago have been found in Egypt that share characteristics with modern platyrrhines, such as three premolars instead of two, implying, perhaps, that the ancestors of New World monkeys started out in Africa (Fleagle 1999; Seiffert 2006). Because fossils of platyrrhines have been found in Bolivia that are about 26 million years old, the founders of the platyrrhine radiation likely dispersed over water to the New World (Fleagle 1999). They could have done this on one or more very large floating islands dislodged from flooding rivers in West Africa sometime after 34 million years ago but earlier than 26 million years ago. They have remained in the New World ever since.

While platyrrhine monkeys diversified in South and Central America, catarrhines remained in the Old World. Catarrhine primates, with their two premolars, show up clearly for the first time in the fossil record in Egypt at about 34–32 million years ago (Rasmussen 2002b; Seiffert 2006). Later, they spread to Europe and Asia but not to North or South America, or Madagascar.

Where Did Primates Originate?

Based on the fossil record, there are two major schools of thought about where primates initially evolved. Some scientists favor an origin in North America or Asia, the landmasses that once made up part of the northern hemisphere's supercontinent of Laurasia (Figure 2.2). North America, in particular, has yielded numerous fossils of primates and mammals called plesiadapiforms, which appeared in the Paleocene epoch and lived between 65 and 50 million years ago (Figure 2.3) (Fleagle 1999). Some paleontologists argue that plesiadapiforms were, in fact, primates and the direct ancestors of the truly modern-looking primates, called euprimates (Figure 2.4) (McKenna and Bell 1997; Bloch and Boyer 2002; Bloch et al. 2007). Others argue that plesiadapiforms were not primates but were more rodent-like (Cartmill 1972, 1974; Maas et al. 1988; Martin 1993). I will return to plesiadapiforms later in this chapter.

Other scientists favor Africa as the landmass of origin (Rasmussen 2002a). That continent, along with South America, Antarctica, Australia, India, Pakistan, and Madagascar, was once part of the southern hemisphere's supercontinent of Gondwana (see Figure 2.2). Potentially in support of an African origin is *Altiatlasius,* a 60-million-year-old creature found in Morocco. It was first described as a euprimate (Sigé et al. 1990), but later analysis suggested that it could be a plesiadapiform (Hooker et al. 1999). The matter is still unresolved (see, e.g., Tabuce et al. 2004). If *Altiatlasius* is not a plesiadapiform, it is figuratively hanging by its teeth as a euprimate because teeth are all that have been found.

Traditional ideas about where primates evolved are based on the fossil record. Although paleontologists may disagree on where primates originated, nearly all assume that primates could not have evolved in Madagascar because no fossil primates have been found there and because it has been isolated from Africa for about 160 million years, long before today's existing orders of placental mammals (called crown-group placental mammals) evolved. The presence of primates in Madagascar today is thought by

Figure 2.2   Laurasia and Gondwana before they split apart into the smaller continents that we recognize today.

Figure 2.3   A reconstruction of *Carpolestes,* a plesiadapiform. Plesiadapiforms are sometimes argued to be primates.

Figure 2.4   A reconstruction of a typical early but extinct primate called a euprimate, which lived 55 million years ago in North America.

most to be the result of travel from Africa or Asia to Madagascar on float-ing islands, much as primates are thought to have arrived in South America from Africa (Simpson 1940; Fleagle 1999; Kappeler 2000). An alternative theory is that prosimians arrived in Madagascar via a temporary land bridge connected to Africa over the Mozambique Channel (McCall 1997).

Fossils and present views notwithstanding, it is important to remember that where primates originated is really without conclusive evidence at this point and is therefore still open to question. The fact that putative pri-mates more ancient than the North American plesiadapiforms have not been found in landmasses that were formerly Gondwana does not neces-sarily mean primates so ancient never existed there. Part of the reason for the lack of fossils in Gondwanan landmasses is that much less investiga-tion into the Cretaceous period and Paleocene epoch has been concen-trated there. Unfortunately, even if we begin to focus more on Gondwana, progress will be frustratingly slow because fossils will be hard to come by. Several of the southern landmasses are highly stable geologically (eastern Africa, with its active volcanoes and frequent rifting, is a well-known exception) (G. Vermeij, pers. comm., 2 June 2008), resulting in conditions that make exposure of suitable deposits rare. While the more geologically active northern continents have yielded a relatively large col-lection of fossils, and not just of primates but also of other animals, the southern landmasses have been rather miserly about giving up their Cre-taceous and Paleocene secrets (Figure 2.5).

Traditional ideas about when primates evolved are also based on the fossil record. For as long as we have recognized that fossils constitute evi-dence of animals that lived a long time ago, we have used them to inform us about where and when they lived. As a result, until very recently it was accepted that not only primates but also all crown-group placental mam-mals evolved in Laurasia, and that they evolved at the boundary of the Cretaceous and the Tertiary periods during the Paleocene epoch, i.e., around 65 million years ago when the dinosaurs went extinct. The fossil record shows modern mammals, including primates, appearing abruptly in the northern continents at the end of the Paleocene and the beginning of the Eocene epochs (about 55 million years ago) without any known pre-cursors (plesiadapiforms notwithstanding). A northern origin at about 65 million years ago was quite reasonable in the absence of any other evi-dence. Now, however, evidence of a different sort is emerging that enables us to consider alternatives to a 65- or 60-million-year-old African, Asian, or North American origin for primates. I want to spend the rest of this chapter discussing one such alternative because it is newer and thus less well known than all the others.

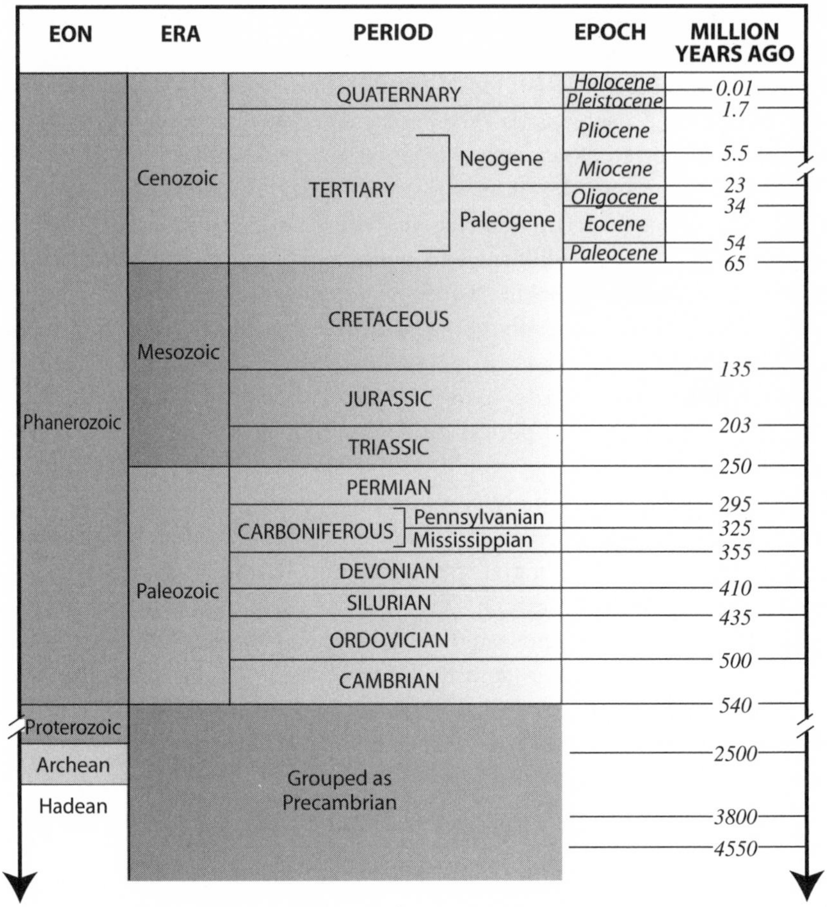

EON	ERA	PERIOD		EPOCH	MILLION YEARS AGO
Phanerozoic	Cenozoic	QUATERNARY		*Holocene*	*0.01*
				Pleistocene	*1.7*
		TERTIARY	Neogene	*Pliocene*	
					5.5
				Miocene	
					23
			Paleogene	*Oligocene*	*34*
				Eocene	
					54
				Paleocene	*65*
	Mesozoic	CRETACEOUS			
					135
		JURASSIC			
					203
		TRIASSIC			
					250
	Paleozoic	PERMIAN			
					295
		CARBONIFEROUS	Pennsylvanian		*325*
			Mississippian		*355*
		DEVONIAN			*410*
		SILURIAN			*435*
		ORDOVICIAN			*500*
		CAMBRIAN			
					540
Proterozoic		Grouped as Precambrian			*2500*
Archean					*3800*
Hadean					*4550*

Figure 2.5   The geological time scale.

Molecular Systematics: An Earlier Origin for Primates?

Molecular biology is a relatively new scientific discipline, but it has grown very quickly. The expansion of molecular biology has included molecular systematics, a field of study that examines ancient relationships by comparing DNA sequences of multiple taxa and analyzing the differences among them. Molecular studies have enabled us to look more closely at the genetic relationships of the various orders of placental mammals, and the implications for primates are intriguing because an earlier date for their origin might open up the possibility of other places of origin.

Importantly, numerous studies using different methods (which I will not discuss here) support a much earlier origin for the ancestors of today's mammals than the fossil record indicates (e.g., Kumar and Hedges 1998; Eizirik et al. 2001; Murphy et al. 2001a; Springer et al. 2004; Bininda-Emonds et al. 2007). Given the rarity of fossils of any kind, an earlier origin should not be surprising; it is highly unlikely that any fossil would represent the first of its kind (Martin 1990, 2000; Tavaré et al. 2002). Most molecular studies point to the origin of crown-group placental mammals at around 108–101 million years ago, well into the Cretaceous period (Kumar and Hedges 1998; Eizirik et al. 2001; Murphy et al. 2001a; Springer et al. 2004; Bininda-Emonds et al. 2007). Molecular evidence also suggests that primates originated about 90–80 million years ago (Martin 2000; Eizirik et al. 2004; Bininda-Emonds et al. 2007) instead of about 65–60 million years ago, as the fossil record implies.

Multiple molecular studies have also consistently supported a revision of the superordinal clades (taxonomic units that include several orders that are genetically related) that were originally based on morphological evidence (Stanhope et al. 1998; Eizirik et al. 2001; Madsen et al. 2001; Murphy et al. 2001a,b; Delsuc et al. 2002; Amrine-Madsen et al. 2003; Springer et al. 2003; Waddell and Shelley 2003; Reyes et al. 2004; Bininda-Emonds et al. 2007; Wildman et al. 2007). One molecularly determined superordinal clade is called Afrotheria, which includes elephants, aardvarks, hyraxes, and tenrecs (small shrew-like mammals), but importantly, not primates. Another superordinal clade is called Xenarthra, which includes anteaters, armadillos, and sloths (Springer et al. 2004). Modern distributions of the animals from these clades suggest that Afrotheria and Xenarthra originated in Africa and South America, respectively. In other words, the only continent that includes elephants, aardvarks, hyraxes, and tenrecs all together is Africa, and the only continent that includes anteaters, armadillos, and sloths all together is South America. It is more parsimonious, that is, less complicated, to infer that the current distribution of these animals is a result of their past distribution than to infer that they all evolved on different continents, found their way to Africa or South America, and then went extinct on their various continents of origin. Scientists go with parsimony in most cases. An origin for crown-group placental mammals on these continents is very much at odds with traditional views based on the fossil record because Africa and South America were Gondwanan landmasses that separated well before crown-group mammals show up in the fossil record (Figure 2.6).

The molecularly estimated dates of the origins of Afrotheria and Xenarthra at 108–101 million years ago actually correspond well with the timing of the geological separation of Africa and South America as a result of plate tectonic activity (Hedges et al. 1996; Eizirik et al. 2001; Murphy et al. 2001a; Springer et al. 2004). Geological evidence suggests that the two continents had finished separating by about 100 million years ago, which would have made it possible for Afrotheria and Xenarthra to diverge and follow their own separate evolutionary paths by vicariance. Evolution by vicariance occurs when animals become separated from each other as a result of geographical separation, e.g., rivers cleaving regions or the movement and isolation of landmasses, not by the movement of the animals themselves, which is called dispersal.

Two other major superordinal clades are Laurasiatheria and Euarchontoglires. Members of Laurasiatheria include bats, carnivores, whales, artiodactyls (e.g., hippos, deer, and antelopes), and perissodactyls (e.g., horses). Members of Euarchontoglires, recall, are restricted to primates, rodents, lagomorphs (rabbits, hares, and pikas) treeshrews, and colugos (Figure 2.7). Some molecular systematists speculate that these two clades diverged in Laurasia when it separated from Gondwana, but it is unclear

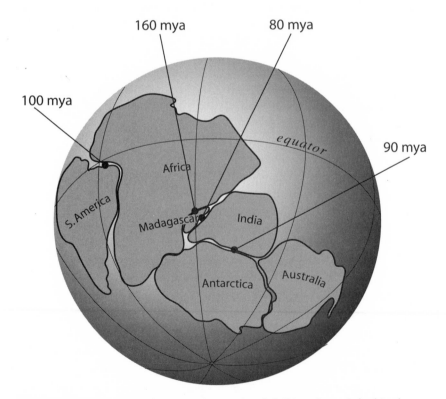

Figure 2.6   The estimated timing of continental drifting that resulted in the breakup of Gondwana; mya = million years ago.

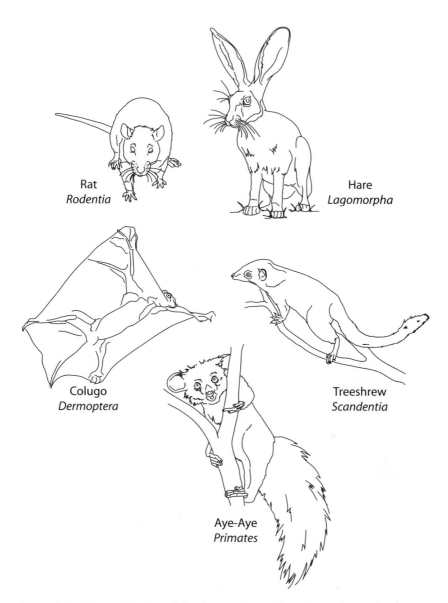

Figure 2.7 Representatives of the five modern orders that make up the
superordinal clade of Euarchontoglires.

how the divergence would have occurred (Wildman et al. 2007). Whereas Xenarthra and Afrotheria were helped along by the plate tectonic activity that separated South America and Africa, North America and Asia were not always completely separated from each other (this explains how early euprimates could exist in both North America and Eurasia).

Other molecular studies estimate that both Laurasiatheria and Euarchontoglires branched off from Xenarthra 100–88 million years ago (Murphy et al. 2001a; Springer et al. 2003). If this was the case, Laurasiatheria and Euarchontoglires (including primates) would have had their beginnings in Gondwana along with the other placental mammals. Because the fossil record makes it clear, however, that representatives of both Laurasiatheria and Euarchontoglires were in Laurasia no later than 55 million years ago, members of Laurasiatheria and Euarchontoglires would have had to disperse to Laurasia earlier than that. Was it possible? And how would they have solved the problem of differentiating on connected landmasses of Laurasia?

Could Indo-Madagascar Be Eden?

It may be possible to answer both questions by building a scenario in which Laurasiatheria and Euarchontoglires evolved into separate clades from a common proto-Xenarthran ancestor (before Xenarthrans themselves became specialized) by dispersing from South America in two different directions. Indeed, two routes for the dispersal of animals from South America to Laurasia have been proposed (Figure 2.8) (Murphy et al. 2001a). One route involves dispersal to North America while it was briefly connected to South America about 90 million years ago. Proto-members of Laurasiatheria might have taken this northern route out of South America (Murphy et al. 2001a).

The other route involves dispersal to the combined landmass of India and Madagascar via Antarctica, which was connected to both South America and Indo-Madagascar perhaps as late as 80 million years ago (Murphy et al. 2001a; Rage 2003; Krause et al. 2006). One potential land bridge, now under water, that may have connected Antarctica and Indo-Madagascar is the Kerguelen Plateau (Figure 2.9). This land bridge is thought to have connected India to Antarctica until about 95–80 million years ago (Storey et al. 1995; Hay et al. 1999; Cracraft 2001; Mohr et al. 2002; Ali and Aitchison 2008). Another land bridge, the Gunnerus Ridge, was suggested to have connected Madagascar to Antarctica until about 82 million years ago (Case 2002; Rage 2003), but more recent evidence argues against that (Ali and Aitchison 2008).

Recent drilling into the Kerguelen Plateau has revealed that ferns, conifers, and, most importantly, early angiosperms (flowering plants) existed there 95 million years ago (Mohr et al. 2002; Frey et al. 2002). These could have existed only in lands that were above water, and, as will be made clear later on in the book, angiosperms may have been crucial to the evolution of primates.

The presence of similar fossil dinosaurs in Madagascar, India, and South America has been attributed to land connections between South America and Indo-Madagascar (Krause et al. 1997, 2006; Sampson et al. 1998). A land bridge connecting Indo-Madagascar with South America via Antarctica also explains how a number of modern and ancient animals in Madagascar have their closest relatives in South America and not Africa (Krause et al. 2006). These include certain species of frogs, lizards, and constricting snakes (Macey et al. 2000; Rogers et al. 2000; Vences et al. 2001; Evans et al. 2008).

This southern route would have enabled animals eventually to reach Laurasia via India after India separated from Madagascar about 80–90

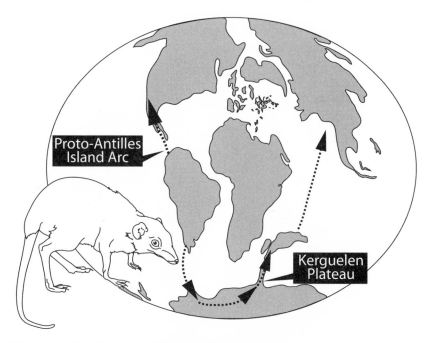

Figure 2.8   Possible land bridges enabling mammals to disperse from South America to North America and to Indo-Madagascar about 80–90 million years ago. The animal shown here represents a generalized common ancestor of Laurasiatherians and Euarchontoglirans.

million years ago and sped northward to establish contact with Asia. The southern route likely would not have been taken by proto-Laurasiatherians because the only non-flying Laurasiatherians in Madagascar today (carnivorans) are not ancient but arrived there less than 25 million years ago (Yoder et al. 2003). More likely, the southern route would have been taken by proto-members of Euarchontoglires (Isbell 2006b), represented in Madagascar today by rodents, which are recent arrivals to the island, and primates, the most ancient "arrivals" (Yoder et al. 1996; Jansa et al. 1999; Poux et al. 2005).

The split between Euarchontoglires and Laurasiatheria is estimated to have occurred about 94 million years ago (Springer et al. 2003), corresponding with both the change from terrestrial to marine paleoenvironments on the Kerguelen Plateau and the severing of a land bridge connecting North and South America (Mohr et al. 2002; Graham 2003; Hedges 2006; Krause et al. 2006), which effectively isolated South America from Indo-Madagascar and North America. The dispersal of animals from South America in different directions, followed by their isolation for millions of years in Laurasia and Indo-Madagascar, could explain the divergence of Laurasiatheria and Euarchontoglires. The severing of India

Figure 2.9 The Kerguelen Plateau may have provided a land route for animals to Indo-Madagascar from South America via Antarctica.

from Madagascar and India's eventual connection with Asia could also explain the rather sudden appearance of Euarchontoglires in Laurasia.

Until now, Madagascar has never been considered the birthplace of any modern placental mammal group because its isolation from Africa occurred about 160 million years ago, well before crown-group placental mammals evolved. But if Indo-Madagascar was connected to South America via Antarctica, and mammals were present 100 million years ago in South America, the possibility certainly arises that mammals were able to disperse to Indo-Madagascar from South America long after Indo-Madagascar separated from Africa. It now becomes conceivable for Indo-Madagascar to have been the birthplace of some of today's orders of mammals, including primates.

An origin of primates in Indo-Madagascar eliminates the need to invoke rafting across water on floating islands, thus solving one particularly problematic puzzle. This is timely because there now seems to be evidence against the rafting idea. A detailed examination of wind patterns, water currents, cyclones, and hurricanes, all of which could have been potential propellers of passive, water-intolerant primates over water, has recently traced the paths that floating islands could have taken from Africa. No path comes close to Madagascar (Stankiewicz et al. 2006).

An Indo-Malagasy origin for primates makes logical sense but it creates three new major questions that need to be considered before it can be taken more seriously.

One question that arises is that if primates originated in Indo-Madagascar, how would they have reached Laurasia in time to deposit their fossils at least 58–55 million years ago? India has long been thought to have collided with Asia about 50–45 million years ago after it separated from Madagascar (Dewey et al. 1989). Obviously it would have been impossible for primates in North America and Asia 55 million years ago to have been descendants of primates from Indo-Madagascar if India collided with Asia close to 10 million years later. Is there any new evidence of earlier contact that can resolve this problem?

The second question is somewhat related to the first. Today Africa is home to prosimian galagos and pottos. Assuming that primates evolved in Indo-Madagascar and that we have dispatched the first problem effectively, the inverted question of how prosimians got from Madagascar to Africa would then need to be resolved.

The final question is that if primates were part of an ancient clade that originated as a result of dispersal and subsequent isolation in Indo-Madagascar, why are the other members of that clade not also in Madagascar today? Actually, as I mentioned earlier, rodents are in

Madagascar, but they are newcomers, apparently arriving from Africa or Asia much later than primates (Michaux and Catzeflis 2000; Poux et al. 2005). Lagomorphs, treeshrews, and colugos, the other members of Euarchontoglires, are not there. Why?

Let us deal with these concerns in order. First, if primates originated in Indo-Madagascar, India would have had to come close to Asia earlier than 55 million years ago because fossil euprimates in North America first show up in rocks that are about 55 million years old (Smith et al. 2006). If we concede that plesiadapiforms were primates then contact would have to have occurred even earlier, sometime before 65 million years ago, because *Pandemonium* (yes, that is the name), the earliest certain plesiadapiform, is estimated to have lived in North America in the early Paleocene, about 65 million years ago (Fleagle 1999). Is there any evidence that India made contact with Asia in time for these fossils to have been deposited in North America?

In fact, there is. In the past decade or so, evidence from numerous scientific disciplines has accumulated that indicates that India was closer to Asia much earlier than has traditionally been thought (Jaeger et al. 1989; Beck et al. 1995; Rage et al. 1995; Briggs 2003; Ali and Aitchison 2008; but see also Thewissen 1990). Importantly, a growing body of research reveals that India and Asia exchanged animals and plants as early as 65 million years ago. There is evidence of an Indian origin for diverse organisms in Asia, including plants (such as figs), insects (such as wasps), cichlid fishes (similar to tilapia), amphibians, chameleons and agama lizards, and ratite birds (e.g., ostriches) (Macey et al. 2000; Bossuyt and Milinkovitch 2001; Cooper et al. 2001; Cracraft 2001; Machado et al. 2001; Conti et al. 2002; Gauld and Wahl 2002; Gower et al. 2002; Wilkinson et al. 2002; Morley and Dick 2003; Rutschmann et al. 2004; Sparks 2004). Dispersal from India to Asia with subsequent divergence has been dubbed the "Out of India" hypothesis. The divergence of a small but diverse set of organisms indeed coincided with the breakup of Gondwanan landmasses and later dispersal to Asia from India in the early Paleocene.

This new estimate of an earlier time of faunal and floral contact between India and Asia thus takes care of one potential problem: it allows sufficient time for euprimates to have begun their fossil record in northern Laurasian continents.

The second problem deals with modern distributions of prosimians. Although only prosimians are found in Madagascar, they are not limited to Madagascar. Pottos and galagos are found in Africa; lorises and tarsiers, in Asia. The "Indo-Malagasy Eden" hypothesis makes it easy to explain the presence of prosimians in Asia as a result of India's docking with that landmass, but how would prosimians have colonized Africa if they originated in Indo-Madagascar? One possibility is that they could have dispersed to Africa by way of Asia after India docked with Asia. Another possibility involves more direct dispersal from India to Africa (Figure 2.10). As India moved northward in the late Cretaceous and early Paleocene, it might have passed quite close to eastern Africa or become connected temporarily to Africa by a land bridge, which is suggested to have been in place as early as 70 million years ago (Briggs 2003; Masters et al. 2005). Evidence of interchange between India and Africa indeed exists, coming mostly from plants in India that have late Cretaceous African affinities (Conti et al. 2002; Briggs 2003; Morley and Dick 2003).

The timing of both floral interchange at about 68 million years ago and an Indo-Africa land connection (estimates range from 70 to 43 million

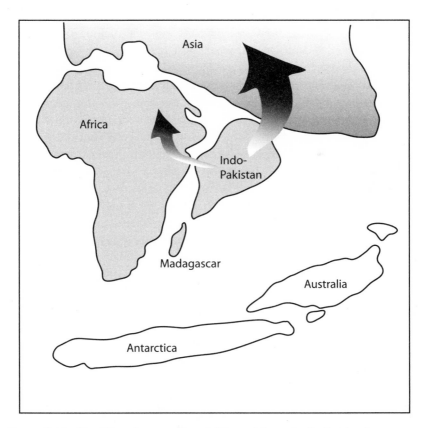

Figure 2.10 Possible primate paths of dispersal from the Indian landmass to Africa and Asia around 65 million years ago.

years ago) (Conti et al. 2002; Roos et al. 2004; Masters et al. 2005) is consistent with the estimated 70–60-million-year-old divergence of galagids and lorisids (the families to which galagos and pottos, respectively, belong) from the Malagasy prosimians (Yoder and Yang 2004). Thus, it is conceivable that prosimians arrived in Africa by jumping the so-called India ship as it passed by. Once there, prosimians would have encountered a continent with an empty niche—primates are not part of Afrotheria and no primates are known to have been present in Africa at that time—making their establishment easier. In the process of becoming isolated in India and then colonizing Africa, prosimians would have diverged from those in Madagascar. The fossil record is also consistent with this scenario. Fossil galagids and lorisids have been found in Egypt and have been dated to the late middle Eocene, about 40 million years ago (Seiffert et al. 2003).

All evidence currently available makes the movement of early primates from India to Africa a reasonable scenario. At the very least, it is no more problematic than their movement from Africa to Madagascar.

This brings us to the final and perhaps most perplexing problem. If primates evolved in Indo-Madagascar, why are their close relatives, the treeshrews, colugos, and lagomorphs, not also in Madagascar? Answering this question requires information about the mammals that are there.

In addition to primates, the only orders of non-aquatic, non-flying mammals represented in Madagascar today are tenrecs, carnivorans (mongoose relatives), and rodents. Molecular work suggests that these mammals arrived in Madagascar at different times, with prosimians being the first of the lot. Tenrecs are members of Afrotheria that invaded Madagascar from Africa 53–37 million years ago (Douady et al. 2002). Carnivorans are members of Laurasiatheria that later spread to Africa and then invaded Madagascar from Africa 24–18 million years ago (Yoder et al. 2003). Although rodents are members of Euarchontoglires, the rodents that exist on Madagascar today are not ancient inhabitants; they appear to have invaded Madagascar 30–15 million years ago (Jansa et al. 1999; Poux et al. 2005). Hippos, the only aquatic mammals known to exist in Madagascar, are now extinct there, but they arrived from Africa in the relatively recent Pleistocene. What strikes me about these taxa in Madagascar is that all except primates have representatives that are aquatic or semi-aquatic on mainland Africa or Asia today, some of whom could have made a successful journey across some extent of water to Madagascar.

In contrast, although lagomorphs are capable of swimming, it is unlikely that any swims well enough (as an unfortunate swamp rabbit's interaction with former president Jimmy Carter's oar reminds us older folks) to envision them being able to make a long and successful over-water journey from Asia or Africa to Madagascar. The distribution of treeshrews, limited today to South and Southeast Asia, and their absence on islands in that same area, indicates that they are also severely limited in their ability to disperse over water (Olson et al. 2005). Finally, there are only two species of colugo and neither is aquatic or semi-aquatic. Colugos are restricted to warm and humid rainforests (Marivaux et al. 2006) and their present distribution on a few oceanic islands could be a consequence of being stranded as ocean levels rose, rather than over-water dispersal.

While a difference in ability to tolerate water might solve the riddle of the current distribution of non-primate mammals in Madagascar, it does not solve a deeper problem. If Euarchontoglires (and primates) originated in Indo-Madagascar, we might expect rodents, lagomorphs, treeshrews, and colugos, fellow clade members with primates, to have also originated in Indo-Madagascar. Is this expectation valid?

In fact, recent molecular findings suggest that although the clade as a whole originated before India and Madagascar separated, only rodents and primates had begun to diversify before the separation (Bininda-Emonds et al. 2007). Rodents and primates appear to have begun their diversification about 88–80 million years ago while India and Madagascar were still connected by land bridges. It is plausible that rodents did exist in Madagascar but that they were subsequently driven to extinction when the more recent mammalian invaders arrived from the mainlands. The subsequent return of different rodent species (as competitors) and the arrival of mongooses (as predators) to Madagascar could have exterminated Madagascar's endemic rodents. Indeed, in modern times we have seen endemic species on other islands exterminated after animals invade from elsewhere (e.g., brown tree snakes [*Boiga irregularis*] in Guam) (Rodda and Savidge 2007). This process of extermination by more recent arrivals is not limited to small islands. For example, extensive replacement of endemic marsupial species in South America by North American mammals is thought to have occurred during the Great American Biotic Interchange after the Panamanian land bridge formed between North and South America about three million years ago (Coates and Obando 1996). If it happened in a landmass as large as South America, it could have happened in much smaller Madagascar.

In contrast to rodents and primates, lagomorphs began their diversification about 67 million years ago, after India separated from Madagascar and around the time India made contact with Asia, whereas colugos began their diversification only about 13 million years ago, long after India and Asia were fully connected. Treeshrews are intermediate, with basal diversification beginning about 32 million years ago (Bininda-Emonds et al. 2007).

The absence of these clade members can be attributed to the fact that they arose in India or Asia only after India and Madagascar separated. Thus, it appears that the third and final problem is indeed surmountable. We are now free to make new predictions based on the new scenario.

One prediction is that the oldest primate fossils in Laurasia should be found eventually in India or, lacking fossiliferous sites there, in Asia. New analyses support this for at least one early euprimate. *Teilhardina* started out in Asia and apparently within about 25,000 years had spread to Europe and North America (Smith et al. 2006). The other prediction is that the oldest fossils of the other orders within Euarchontoglires should be found in Asia, if not India, because either India or Asia is their place of origin and site of initial diversification. So far this is indeed the case: the oldest treeshrews, rodents, lagomorphs, and colugos have all been found in either India or Asia (Meng 2004; Sargis 2004; Asher et al. 2005; Marivaux et al. 2006; Rose et al. 2008).

Were Plesiadapiforms Ancestral to Primates?

Given how quickly the euprimate *Teilhardina* appears to have spread from Asia to North America and Europe, the revised estimate of floral and faunal contact between India and Asia at about 65 million years ago makes it just possible for plesiadapiforms to have arrived in Asia and dispersed to North America by the time (65–62 million years ago) that *Pandemonium,* the oldest certain plesiadapiform, lived in North America (Fleagle 1999). It may be stretching it, however, to think that plesiadapiforms originated in Indo-Madagascar and thus were immediate ancestors of primates. The main issue is that a dispersal route of plesiadapiforms from Asia to North America is inconsistent with evidence that they originated in North America. The many fossil plesiadapiforms in North America and the very few fossils of them in Asia are one indication that they started out in North America. The relative number of fossils by region may, however, merely reflect a bias toward more heavily inspected areas, such as North America. More satisfying evidence, perhaps, is that comparisons of plesiadapiforms in Asia and North America reveal morphological changes that are also supportive of a dispersal direction from North America to Asia (Smith et al. 2004).

If plesiadapiforms were primates, why would they buck the trend followed by the other members of Euarchontoglires and become established first in North America rather than Asia? This unique dispersal direction argues against the idea that plesiadapiforms were primates, or even ancestral to primates. The conclusion that might be drawn from this is that they were not members of Euarchontoglires but rather members of Laurasiatheria. The initial establishment of plesiadapiforms in North America is also consistent with the idea that Laurasiatheria took the northern route out of South America and into North America where they began to diversify.

The timing of the arrival of plesiadapiforms and rodents in North America is also compatible with the idea that plesiadapiforms were members of Laurasiatheria. The arrival of *Pandemonium* in North America about 65 million years ago preceded that of rodents; fossils of rodents in North America have been found in rocks dated to the Eocene, after the plesiadapiforms appeared (Maas et al. 1988). Fossils of plesiadapiforms and rodents in the same location (Bloch and Boyer 2002) also reveal that they coexisted for a time. There is evidence that plesiadapiforms and rodents were so similar that they competed with each other for resources. Indeed, competition from rodents has been suggested as the cause of the extinction of some of the plesiadapiforms (Maas et al. 1988).

Carpolestes was one of those plesiadapiforms that could have been exterminated by rodents (Maas et al. 1988). Like some species of rodents, it had manual and pedal grasping but not leaping, a nail instead of a claw on its hallux (big toe), and no postorbital bar (Bloch and Boyer 2002). Some rodents are quite capable of reaching and grasping and they have good manual dexterity. Their ability to reach for and manipulate objects with the digits of their forepaws has led some to suggest that reaching and grasping evolved even before the rodent and primate lineages diverged (Whishaw et al. 1998; Iwaniuk and Whishaw 2000; Whishaw 2003). While the presence of nails and grasping hands in *Carpolestes* and other plesiadapiforms may reveal little about the evolution of primate characteristics, they may shed light on the characteristics of the common ancestor of Laurasiatheria and Euarchontoglires before those taxa dispersed from South America. The apparent ecological similarity and evolutionary convergence between plesiadapiforms and rodents also suggests that many living rodents might be good models for reconstructing the lives of plesiadapiforms.

Where and when was the scientific equivalent of Eden, the birthplace of primates? With the development of molecular systematics, the choices have broadened. Many paleontologists will continue to argue that, based entirely on the fossil record, Eden was North America or Asia or Africa about 65 million years ago. That position makes sense; fossils are their livelihood. But others may be willing to incorporate molecular work and begin to investigate the possibility that Eden was Indo-Madagascar about 90 million years ago.

Our state of knowledge about Indo-Madagascar as the possible birthplace of primates is reminiscent of the conditions Darwin faced when he predicted Africa as the birthplace of humans. In the absence of any hominin fossils, and going only on the logic of similarities in morphology, Darwin suggested that we humans have our roots in Africa. But, when the first hominin fossils were discovered, they were found in Asia, not Africa, and it became widely believed that Darwin was wrong about our birthplace. As time passed, however, and hominin fossils were discovered in Africa, an African origin was reconsidered and then accepted. Molecular studies have also helped cement Africa as the continent of our origin by confirming what Darwin anticipated—our closest relatives are the African apes not the Asian apes. Our current views of primate origins may be similar to views of hominin origins in the doubting years shortly after Darwin's time. Using fossils alone, the evidence points to an origin in North America, Africa, or Asia. Keep in mind, however, that the fossil evidence is not so strong that any of those regions can be eliminated and that the absence of fossils is sometimes misleading. Although no fossil primates have been found in Madagascar (indeed, nor have many fossils of any kind been found there), the fossil record does not exclude Indo-Madagascar as the birthplace of primates. This, coupled with logic strengthened by indirect evidence, suggests that, to North America, Asia, and Africa, we now need to add Indo-Madagascar as a potential place of origin for primates.

Before I close this chapter, I want to share a bit of the human side of the process of hypothesis-building. As so often happens in science, the information to put together the Indo-Malagasy Eden hypothesis was all out there, available for anyone to assemble the threads. It turns out that at least two people have done that, and done it independently of one another.

My own trajectory toward the Indo-Malagasy Eden hypothesis began in the early days with my Snake Detection theory. I actually included it in the first version of my Snake Detection manuscript but removed it when it generated strong negative emotions from the reviewers. Once that manuscript was finally accepted (which took nearly two years and *four* rounds of reviewer comments and revisions), I returned to the Indo-Malagasy Eden hypothesis and presented a talk about it to the International Primatological Society in Entebbe, Uganda (Isbell 2006b). In the meantime, Robert Martin, a well-known and highly respected comparative primatologist who has studied both living and fossil representatives, had been thinking along the same line (e.g., Martin 2003, 2006). Fortunately, he had more success in publishing the idea than I, and in 2005 he (along with colleagues Ellen Miller and Gregg Gunnell) published a discussion of the Indo-Malagasy Eden hypothesis.

> An Indo-Madagascar origin hypothesis is compatible with all known molecular and morphological evidence involving primate . . . evolution, but too little information is available from the fossil record to test any specific aspect of this hypothesis. However, with regard to biogeography, it appears that the more that is learned about Indo-Madagascar, the more likely it seems that the role of the Indian land mass in scenarios of early primate evolution and dispersal has been underestimated. (Miller et al. 2005: 87)

Although our focus is a bit different (Martin concentrates more on the possibility of an earlier date of origin [e.g., Martin et al. 2007], while I work out the details of how primates could have originated on Indo-Madagascar), the fact that we both arrived at the same hypothesis independent of one another reinforces in my mind that the idea is truly worth pursuing.

CHAPTER THREE

Why Did Primates Evolve?

How often have I said to you that when you have eliminated the impossible, whatever remains, *however improbable*, must be the truth?

Arthur Conan Doyle, *The Sign of Four* (1890: 93)

W HAT I have told you so far is that primates are different from other mammals by virtue of their emphasis on vision as their primary sense, partly expressed by having a high degree of orbital convergence and tight coordination between eyes and hands for reaching and grasping. At some point in the evolutionary history of the Euarchontoglires, at least one of its members was different from the rest, and this difference led that animal and its descendants down a path that eventually resulted in hundreds of species of mammals, now called primates, all having expanded visual abilities. Why have primates, among all mammals, specialized in vision as the primary sense? How did they live? What were their main environmental challenges? What did the ancestors of primates do so differently from other mammals that their descendants became unique among mammals?

These questions have been debated for quite a long time (e.g., Napier and Walker 1967; Cartmill 1974; Sussman 1991; Crompton 1995; Bloch and Boyer 2002; Rasmussen 2002a; Kirk et al. 2003), and there is still no general agreement. Many hypotheses for the origin of primates have been proposed over the years. Most have focused on reconstructing a lifestyle that included a unique diet or a unique way that proto-primates moved about in their environment. These include the Arboreal theory, the (Nocturnal) Visual Predation hypothesis, the Angiosperm/Omnivore hypothesis, and the Camouflage-Breaking hypothesis.

I will describe these hypotheses in chronological order so that you can get a sense of the scientific community thought processes and our expanding knowledge over time, not only of primates but also of their sur-

Table 3.1 Summary of existing hypotheses for the origin of primates.

Hypothesis	Proposer	Date Proposed	Selective Pressure
Arboreal theory	Grafton Elliot Smith (and others)	Early 1900s	Arboreal lifestyle
Visual predation	Matt Cartmill	1972	Insectivorous diet
Angiosperm/ omnivore	Robert Sussman	1991	Frugivorous/insectivorous diet
Nocturnal visual predation	Matt Cartmill	1992	Nocturnal, insectivorous lifestyle
Camouflage-breaking	Robin Crompton	1995	Arboreal lifestyle

roundings. Understanding the environments in which animals live is necessary for understanding the selective pressures that might have been operating during their evolution. The various hypotheses are summarized in Table 3.1.

The Arboreal Theory

The first hypothesis proposed to explain the evolution of primates has been labeled the "Arboreal theory." It was developed in the early 1900s by Grafton Elliot Smith and Frederic Wood Jones, and extended later by W. E. Le Gros Clark (Elliot Smith 1912; Jones 1916; Le Gros Clark 1959). Ancestral primates were thought at the time to have lived in the canopies of tropical trees. According to the Arboreal theory, the sense of smell is not particularly useful in the trees and so primates lost much of their olfactory ability. It was replaced by expansion of the visual system, which, in the three-dimensional, complex environment of tropical forests, also required coordination of hands with eyes to maneuver along the branches. Because natural selection operates through survival and reproduction, it made sense that in animals often living 30 meters (100 feet) or more up in the canopy, natural selection would favor those characteristics that enabled individuals to move among the branches without falling and killing themselves. The proposers of this hypothesis reasoned that the ability to reach for and grasp onto branches would have been a particularly useful skill to have in an arboreal environment, while the necessity of judging distances between branches before leaping would have favored bringing the eyes closer together toward the front of the face.

The (Nocturnal) Visual Predation Hypothesis

The Arboreal theory stood virtually unopposed for about 50 years until Matt Cartmill (1972) challenged it by looking beyond primates to other mammals that fill the arboreal niche. He pointed out that there are many arboreal mammals that survive and reproduce quite well without grasping hands, nails, orbital convergence, and forward-facing eyes. Tree squirrels are one familiar example. Cartmill also pointed out that many arboreal mammals still retain excellent olfactory capability, and so life in the trees cannot alone explain primates' weakened olfactory sense. Again using a comparative approach, Cartmill proposed an alternative model he originally called the Visual Predation hypothesis. Cartmill proposed that stalking and grabbing insects at close range while on small-diameter branches at lower levels of tropical forests favored the entire suite of primate characteristics (Cartmill 1972, 1974, 1992). Small animals with grasping hands and no claws would have a hard time staying on thick, heavily sloping branches, but they could very easily grasp onto fine, terminal branches. Because other animals that have forward-facing eyes are predators, and because many prosimians today are insectivorous predators, Cartmill reasoned that the first primates were also insectivorous predators. This hypothesis has been highly appealing for over 30 years partly because it skillfully weakened the Arboreal theory and partly because it was supported by carefully developed arguments and lots of examples drawn from other animals.

One such example is that fossil evidence shows that many early primates were small, and because many small modern mammals are insectivorous, it is likely that early primates were also insectivorous. Another example is that while many arboreal mammals do not have forward-facing eyes, some terrestrial mammals that do, such as cats, are considered visual predators (Cartmill 1972, 1974). This also extends to predatory birds. Cartmill used owls and hawks as examples of avian visual predators with forward-facing eyes.

One of the early problems with the Visual Predation hypothesis was that not all predators have forward-facing eyes. Mongooses are one such example of a carnivore with more laterally placed eyes. To account for this, in 1992 Cartmill modified his hypothesis by stating that only those visual predators that are nocturnal need orbital convergence and forward-facing eyes in order to see clearly that which lies in front of them. Diurnal predators do not need to have forward-facing eyes because their pupils constrict in bright light and serve to sharpen the image so that they are still able to see clearly even with laterally placed eyes.

Two recent studies support the (Nocturnal) Visual Predation hypothesis. A field study of slender lorises *(Loris lydekkerianus)*, nocturnal prosimians that have nicely convergent orbits, confirmed that they primarily use vision to locate and capture arthropods, their main food type (Nekaris 2005). Another study confirmed that nocturnal predatory mammals have more convergent orbits than their diurnal predatory relatives, as the (Nocturnal) Visual Predation hypothesis predicts (Ravosa and Savakova 2004). Ravosa and Savakova (2004) also found that early euprimates had orbital convergences similar to those of felids (which are largely nocturnal) and living nocturnal insectivorous primates. The implication is that early euprimates were both nocturnal and insectivorous.

The Angiosperm/Omnivore Hypothesis

The (Nocturnal) Visual Predation hypothesis was challenged in 1991 by Robert Sussman (Sussman 1991) who pointed out that most of the smaller primates, while eating insects to some degree, also eat a lot of plant material. He also pointed out that nocturnal prosimians typically locate their insect prey using their sense of smell or hearing. Their ears orient toward insect sounds, and they are even able to follow moving insects in experimental conditions when barriers are erected to block them from seeing the insects. This is supported in part by a recent study of mouse lemurs *(Microcebus murinus)*. Mouse lemurs were not able to detect or capture non-moving simulated arthropods but could detect and capture them if they moved or were associated with rustling sounds (Siemers et al. 2007). Sussman claimed that other mammalian nocturnal predators also use hearing to catch their prey. For instance, cats that are blindfolded can still catch mice. Finally, he pointed out that Old World fruit bats are the mammals that most closely resemble primates in their visual systems and, as their name tells us, they eat fruits, not insects. Sussman had no problem with the idea that grasping hands evolved for dealing with life on fine, terminal branches but he concluded that (nocturnal) visual predation could not explain the unique visual characteristics of primates.

As an alternative to the Visual Predation hypothesis, Sussman suggested that the first primates were not committed insectivores but were omnivores, primarily eating fruits and other plant foods while taking insects more opportunistically (see also Richard 1985; Rasmussen 2002a). Sussman hypothesized that the first primates were able to take advantage of the appearance of angiosperms (flowering plants), which had begun to spread throughout the world. Angiosperms today include grasses, herbs, small shrubs, and enormous tropical trees, but he suggested that the first primates lived and ate among early angiosperms, which were small shrubs with small-diameter branches. The new foods offered by angiosperms included fruits and flowers, many of which were small and located in the dim light of forest understories. He argued that grasping hands and visual adaptations would be advantageous for primates under those conditions.

In support of Sussman's hypothesis, a morphological comparison of fossil euprimate teeth with those of living prosimians suggests that some early euprimates did eat fruit to some degree (Covert 2002). The shape of primate molars is a good indicator of their diet. High, sharp crests are good for cutting through leaves or the exoskeletons of insects, whereas low, rather flat surfaces are good for eating softer foods such as fleshy fruits. Although we do not have a morphological identifier of flower and nectar eaters, there is intriguing ecological evidence from modern primates that early euprimates might have also consumed nectar. Plants entice pollinators to visit by producing nectar for them to eat. While insects are the most numerous plant pollinators, some mammals are also effective plant pollinators. Intriguingly, prosimians, the living primates considered closest in morphology to early euprimates, represent the majority of mammalian plant pollinators (Carthew and Goldingay 1997). For example, the black and white ruffed lemur, a rather large prosimian in Madagascar, is one of very few animal species that pollinate the traveler's tree *(Ravenala madagascariensis)*. This relationship could be very old. *Ravenala* was among the first genera of a family (Strelitziaceae) that began to diversify in the late Cretaceous and early Tertiary in Madagascar, about the time that molecular studies indicate that primates began to diversify (Kress 1993; Kress et al. 1994).

A study initially designed to test between the (Nocturnal) Visual Predation and Angiosperm/Omnivore hypotheses revealed, however, that diet seems to make little difference in orbital convergence. Woolly possums *(Caluromys)* are nocturnal marsupials from South America that have rather large eyes and brains, and digital dexterity comparable to that of prosimians (Rasmussen 2002a), but little orbital convergence (Sussman 1991). Woolly possums were found to eat both fruits and insects at the tips of fine, terminal branches.

The Camouflage-Breaking Hypothesis

The (Nocturnal) Visual Predation hypothesis was also challenged by Robin Crompton (Crompton 1995). Among other criticisms, Crompton argued that orbital convergence is not terribly important for providing depth perception because shading, perspective, and other cues can inform about depth. Work testing the use of just one eye suggests that natural scenes do indeed provide extensive cues for determining depth (Watt and Bradshaw 2000). On the other hand, when speed and accuracy are important, two eyes do make a difference (Watt and Bradshaw 2000; Melmoth and Grant 2006).

Crompton argued that orbital convergence and the stereoscopic vision it provides more crucially enable animals to break through camouflage. Although he supported the Angiosperm/Omnivore hypothesis by arguing that the early fruits of angiosperms would have been very small and inconspicuous, especially against a complex backdrop of leaves, he did not limit the benefits of orbital convergence to finding camouflaged food. Crompton argued that orbital convergence would be useful for discriminating between any number of small targets. For arboreal animals, such targets would also include branches used during locomotion. He thus offered what I call here the Camouflage-Breaking hypothesis. He argued that orbital convergence and grasping hands would have been useful not only for capturing insects and finding and eating small fruits but also for aiming for and leaping to small branches in the complex three-dimensional environments that are typical of tropical forests.

In support of this hypothesis, morphological evidence suggests that leaping was a common mode of locomotion in early euprimates, as it is in modern prosimians. Many prosimians, the primates most similar to euprimates in morphology, have a locomotor style called vertical clinging and leaping whereby they move by leaping from tree trunk to tree trunk (Napier and Walker 1967; Dagosto 1988; Crompton 1995; Covert 2002; Gebo 2002). Crompton suggested that leaping, although inefficient, was also a safe way to move about because predators cannot easily predict where animals that leap will end up.

A key issue for the future of this hypothesis is to determine to what extent ancestral euprimates leapt. It also remains to be seen just how well eyes and hands coordinate with each other when vertical clingers and leapers leap. I have seen a number of Malagasy lemurs leap, and their feet invariably land first, not their hands. Nor do they watch their hands grasp onto trunks and branches. Finally, by the time they land, they are already looking ahead to the next landing place. On the other hand, such forward-looking behavior may provide just enough hand-eye coordination to allow them to make solid landings.

Do All Primates Have Visually Guided Reaching and Grasping?

All these hypotheses link the evolution in primates of orbital convergence and forward-facing eyes with the evolution of visually guided reaching and grasping. Is this ability characteristic of all primates? If prosimians today differ in the ability to coordinate their eyes with reaching or grasping, it would no longer be possible to infer that all euprimates had visually guided reaching and grasping, and it would be a major blow to the hypotheses just presented.

The only study done so far, and this was over 40 years ago, hints that there might indeed be variation in the degree to which vision assists prosimians in reaching and grasping (Bishop 1964). In that study, lorises, galagos, and lemurs were tested to determine how they used their hands to reach and grasp food. Unfortunately, the study was not conducted to address primate origins and so used different procedures for the subjects. Still, the results are informative. Lorises *(Loris tardigradus)* do appear to use their vision to judge the location of food. One regularly attempted to catch moving mealworms that were behind a transparent baffle, but required extensive training to search for mealworms that it could not see. Lorises' reliance on vision has since been confirmed with observations of natural foraging behavior in the wild (Nekaris 2005). In contrast, galagos *(Galago senegalensis)* appear unable to visually adjust their reaches to moving targets. When mealworms were put on a turntable and spun around at different speeds, the galagos were unable to catch them. The galagos misjudged and struck where the mealworms had been when they first started their reach and they were unable to adjust once they initiated their movement. Galagos normally manage to catch moving prey by using very fast arm movements, which compensate for their inability to make visually guided corrections. Finally, the various lemurs often did not use their hands to reach for and grasp food but obtained the food directly with their mouths. Thus, it is unclear whether lemurs have visually guided reaching and grasping, at least for feeding. This study hints that there may be some variability in arm-hand-eye coordination of feeding among prosimians, and it would be informative to test this more systematically.

If prosimians are found not to differ in visually guided arm and hand movements, we still need to ask if prosimians and anthropoids differ in that regard. Recently, a neurological study of galagos *(Otolemur garnetti)* revealed that stimulation of particular subregions in the posterior parietal cortex (PPC) elicited different responses of the arms and hands. One subregion was involved in movement resembling feeding behavior (hand-to-mouth movements) and another was involved in reaching and grasping movements. Interestingly, these regions receive little input from visual areas (Stepniewska et al. 2005). In contrast, regions for reaching and grasping also exist in the PPC of macaques that do have input from visual areas (Stepniewska et al. 2005), suggesting that there might well be a difference between prosimians and anthropoids in the extent to which vision is used in reaching and grasping.

It is a bit surprising that at this advanced stage of our knowledge of primate behavior, we can still wonder about the universality of visually guided reaching and grasping when it is such an important part of all current hypotheses on the origin of primates. It also seems that if we are ever to understand why primates evolved, we need to incorporate neuro-scientific research into our hypotheses because so much of vision is neurological. The difference between galagos and macaques in visual connections to areas in the brain that control reaching and grasping is a good example of the potential power of neuroscience to inform evolutionary theory.

CHAPTER FOUR

Primate Vision

The brain, like the power plant, can never be shut down and funda-mentally reconfigured, even between generations. All the old con-trol systems must remain in place, and new ones with additional capacities are added on and integrated in such a way as to enhance survival.

John Allman, *Evolving Brains* (1999: 41)

ALL ANIMALS must be able to find food and they must also avoid being killed by others. This is life. In the process of living, ani-mals use senses of smell, hearing, touch, and sight to varying degrees. Eyes first appeared about 543 million years ago during the Cam-brian, when they evolved in trilobites (Parker 2003). This time period is associated with an extraordinary fossil record of life over a span of about five million years and it has been called, fittingly, the Cambrian Explo-sion. At that time, all the major phyla of animals that exist today ap-peared on the scene, and in the following years they evolved into life forms as diverse as flatworms, octopuses, insects, and birds. Vertebrates, a group of animals with backbones that belong in the phylum Chordata, first appeared about 485 million years ago (Figure 4.1).

There are numerous hypotheses about the causes of the Cambrian Explosion, but most have flaws that render them untenable. Andrew Parker, a paleontologist from Oxford University, has recently pre-sented an intriguing theory in which he proposes that the Cambrian Explosion was precipitated by the abrupt evolution of vision in re-sponse to life lived in the light (Parker 2003). According to Parker's "Light Switch" theory, once an atmospheric change increased the amount of light reaching the earth, the benefits of seeing were so over-whelming that eyes not only quickly evolved but they also evolved in several lineages independently. Eyes led to active predation whereby animals were able to see and therefore pursue and kill their prey. The evolution of hard body parts such as teeth and jaws quickly followed to assist in active predation, and, in response, other hard body parts such as shells evolved to protect smaller animals from being eaten. The result was an explosive diversification of life, the likes of which have never been repeated in the history of the earth. If the Light Switch theory is correct, it means that vision has always been associated with predator-prey interactions.

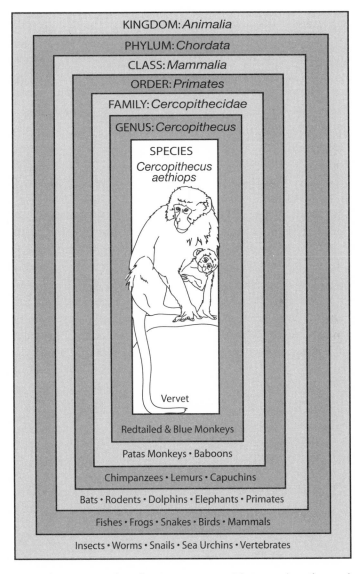

Figure 4.1   The Linnean classification system, with examples of nested groupings.

It is not just eyes, however, that help many animals see. Vertebrate eyes have numerous structures in the brain to support them. Working together, eyes and the specific parts of the brain that enable animals to see in the light are what create visual systems. When vertebrates first arrived about 485 million years ago, they had eyes. Since then, vertebrates have diversified into many different shapes with many different lifestyles. Horses, dogs, whales, rats, bats, and monkeys are familiar to us all, but they all represent only the mammals, one small class of vertebrates. Fishes, frogs, dinosaurs, birds, snakes, and lizards are also vertebrates. Because of their common ancestry, all vertebrates today have a similar simple eye design with a lens and a retina (insects have independently evolved a different sort of eye, the compound eye). All vertebrates also share a few basic brain structures that support the eyes. Primates, however, have modified those structures extensively compared with other mammals.

If we are to understand why primates are the only order of mammals to specialize in vision as their main sense, and thus how and why they evolved, we first need to understand their vision and how it differs from other mammals. What follows is a very brief summary of the neuroscience of mammalian vision, with emphasis not only on the distinctions between the visual systems of primates and other mammals but also among those of prosimians, platyrrhines, and catarrhines, because even in primates, vision is not all the same. Two caveats: first, this summary cannot do justice to the volume of work that has gone into improving our understanding of vision, and I hope that the many experts whose work is not recognized here will forgive me. Second, in my own exploration of neuroscience, my understanding was often painfully slowed by scientists' heavy use of jargon and acronyms. I do not wish the same for you; I will try to write in plain English, but I am afraid I must use some terms that are not yet part of our daily speech. Just take it slowly and look frequently at the figures.

Three of the terms that you will be reading about need to be made very obvious from the beginning. I use *visual system* to refer to all the parts of the brain that give us the ability to see. It is the broadest term of all, but it can be separated into the lateral geniculate nucleus (LGN) visual system and the superior colliculus–pulvinar (SC-pulvinar) visual system. I use *pathway* to describe a distinct set of neural connections going from the eye to various other parts of the brain. There are three of them: the magnocellular (M) pathway, the parvocellular (P) pathway, and the koniocellular (K) pathway. Finally, I use *stream* to describe the conceptual idea that several functions related to vision in the primate brain, e.g., object recognition/assessment and spatial localization/self-movement, can be separated to some degree. The former involves the ventral visual processing stream while the latter involves the dorsal visual processing stream. Other terms will be explained as we approach them. The appendix at the end of the book provides functional definitions for easy reference.

Eyes

Although all mammals have a retina in each eye, the primate retina is a bit different. Only primates (but not all prosimians) have a retinal fovea. This is a pit, or depression, in the retina that allows clear central vision for distinguishing between exceedingly small objects, such as individual needles on a Monterey pine tree decorated with balls and lights in the middle of one's living room, the objects in my view as I write this. In vertebrates, visual information is relayed to two major visual systems by way of nerve fibers, or axons, from ganglion cells in the retina that travel to visual structures.

The Lateral Geniculate Nucleus Visual System

One of those structures is called the dorsal lateral optic nucleus in reptiles and birds and the lateral geniculate nucleus (LGN) in mammals (Butler 1994). The LGN is located in the forebrain's thalamus, and is part of the LGN visual system (Figure 4.2). Primate LGN visual systems are clearly more expansive and more complex than those of other mammals. This is exemplified by primate LGNs themselves, which are larger and more complex in their cellular organization than those of other mammals. Most rodents, for example, have simple LGNs with no obvious layering of cells (squirrels are an exception), whereas cats have LGNs with some cellular layering (Jones 1985). Primate LGNs are traditionally considered to have up to six cellular layers (Figure 4.3). Even among primates, however, there is variation in this layering.

Mammals with differentiated LGNs, such as cats and primates, have visual pathways that are labeled according to the LGN layer to which ganglion cells from the retina travel. The terminology is different for primates, adding to the confusion for non-specialists. Primates are currently thought to have three visual pathways; I will discuss two of the three now and save the third for another chapter. They each have different functions.

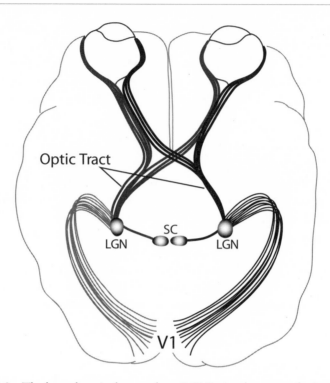

Figure 4.2 The lateral geniculate nucleus (LGN) visual system of primates, exemplified by a human brain. The LGN receives input from the retina via the optic tract. Note the extensive neural projections (the optic radiations) from the LGN to the primary visual area (V1). The superior colliculus (SC) is another recipient of retinal input.

The M Pathway: Vision for Motion Detection

In primates, one visual pathway of the LGN visual system is called the magnocellular (M) pathway, so named because the M retinal ganglion cells, also called parasol cells, project to very large (magno) cells in the LGN; in non-primate mammals, the equivalent pathway is called the Y pathway. In primates, the M retinal ganglion cells project to what have been traditionally counted as the two bottom, M, layers of the LGN. The LGNs of all primates have two distinct M layers (Kaas and Huerta 1988). The M/Y pathway appears to be conservative in its functions across the mammals that have been studied; its neurons are highly sensitive to motion and contrasts in illumination (Jones 1985; Allman and McGuinness 1988; Kaas and Huerta 1988). Because there seems to have been little expansion or modification of the M pathway in primates, I

Catarrhine primates

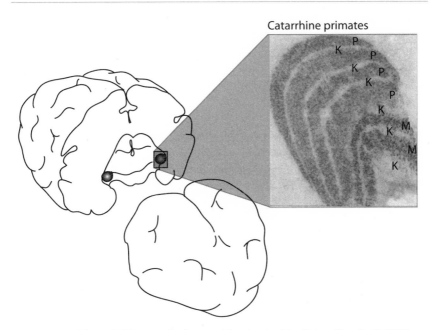

Figure 4.3   The LGN in a typical catarrhine brain. Traditionally, the LGN has been divided into six layers. Layers 1 and 2 are the magnocellular (M) layers. Layers 3 through 6 are the parvocellular (P) layers. More recently, the number of layers has expanded to recognize six koniocellular (K) layers that are situated between the P and M layers.

will not spend any more time discussing it other than to say that it is incorporated into the dorsal visual processing stream, one of two visual processing streams in the brains of primates (Figure 4.4). Visual processing streams appear to be unique to primates because of the great increase in number of areas in the primate brain that are involved in vision. The dorsal visual processing stream goes to the top of the brain to the parietal lobe and includes visual areas V1 and V2, the middle temporal area (MT), and the posterior parietal cortex (PPC) (DeYoe and Van Essen 1988). The dorsal stream is specialized for spatial discrimination, movement detection, and visual control of reaching and grasping (Previc 1990; Goodale and Milner 1992; Ganel and Goodale 2003; Goodale and Westwood 2004). The dorsal stream has been described as being specialized for "vision for action" (Goodale and Milner 1992; Goodale and Westwood 2004). It is also used in perception, however, particularly of spatial relations (Rizzolatti and Matelli 2003).

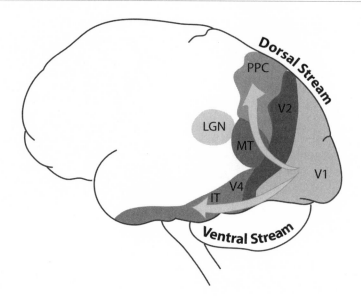

Figure 4.4   Dorsal and ventral visual processing streams. Projections go from the LGN to V1 and beyond to V2, MT, and the posterior parietal cortex (PPC), following the dorsal visual processing stream and to V2, V4, and the inferotemporal cortex (IT), following the ventral visual processing stream. The dorsal processing stream is closely associated with movement, visual control of reaching and grasping, and spatial perception whereas the ventral processing stream is closely associated with object recognition.

The P Pathway: Vision for Object Perception

The second visual pathway of the LGN visual system is the parvocellular (P) pathway; in non-primates its equivalent is the X pathway. In primates the P retinal ganglion cells, also called midget cells, project to the P layers of the LGN. The number and complexity of P layers vary in primates (Figure 4.5). These differences in the complexity of the LGN reveal that the P pathway has not only expanded more in primates than in other mammals but also has expanded more in anthropoid primates than in prosimians, and more in catarrhines than platyrrhines. The P pathway is largely responsible for our own excellent central vision, fine visual acuity, and our ability to see rich color, all of which help us to perceive objects in our environment (Kaas and Huerta 1988). The great expansion of the P pathway in primates clearly requires an evolutionary explanation.

The P pathway is incorporated into the ventral visual processing stream. The ventral stream shares visual areas V1 and V2 with the dorsal stream but then diverges and goes to visual area V4 and the inferotem-

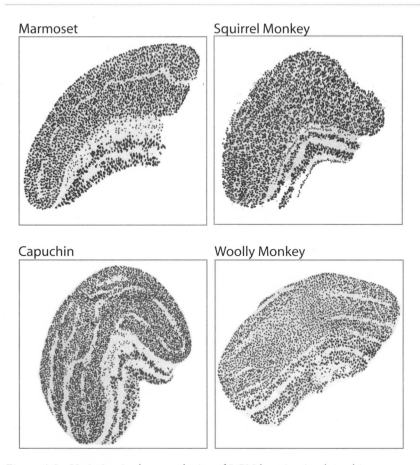

Figure 4.5   Variation in the complexity of LGN layering in platyrrhine primates. New World monkeys (shown here) are more variable and tend to have less complex P layers than Old World monkeys and apes. Adapted from Le Gros Clark (1941).

poral cortex (IT) in the temporal lobe (Maunsell 1987; Maunsell and Newsome 1987; DeYoe and Van Essen 1988) (see Figure 4.4). Early in the ventral stream (V1 and V2), neurons are especially responsive to simple cues of color or lines of different orientations; further along the ventral stream (V4 and IT), neurons become more responsive to combinations of cues that create complex forms (Hubel and Livingstone 1987; Kaas and Huerta 1988; Tanaka et al. 1991; Kobatake and Tanaka 1994). We even have single neurons in the ventral stream that fire in response to images or abstract representations (i.e., written names) of familiar individuals, such as certain politicians and actors (Quiroga et al.

2005). The ventral stream is specialized for scanning, scrutinizing, and recognizing objects, particularly in the central visual field (so it is closely associated with the fovea) (Previc 1990; Rosenbluth and Allman 2002). The ventral stream has been described as emphasizing "vision for perception" (Goodale and Milner 1992; Goodale and Westwood 2004) but I am going to describe it here as "vision for assessment" to distinguish it more from the functions of the dorsal stream, which is also involved in perception (of space).

V1

Once neurons leave the LGN, those in the LGN visual system form what are called optic radiations on their way to visual area V1, which has been variously called Brodmann's area 17, striate cortex, and the primary visual area (talk about confusing!). I will stick with V1 here. All mammals have a V1, which is located in the back of the brain in the occipital lobe. The V1 is a larger and more crucial part of the brain in primates than in other mammals. If a non-primate mammal has a damaged V1, it still has decent visual capability. A cat or a squirrel, for example, can still maneuver around obstacles even without a V1, but if an anthropoid primate loses its V1, it will have severely reduced visual capability (Henry and Vidyasagar 1991; Bullier et al. 1994; Van Hooser and Nelson 2006). Prosimians seem to be intermediate; if a prosimian loses its V1, its visual capability is poorer than a V1-less cat but apparently not as poor as a V1-less anthropoid (Henry and Vidyasagar 1991). In primates V1 is an essential part of the LGN visual system.

In primates most projections go from the LGN to V1, then to V2, and then beyond to other areas of the neocortex. Projections from the LGN can also bypass V1 to go directly to other areas called *extrastriate areas*. Extrastriate projections from the LGN are more common in non-primate mammals than in primates (Henry and Vidyasagar 1991). This is another indication that the LGN visual systems of primates differ from those of other mammals and it may explain the better performance of non-primates with damaged V1s.

The Superior Colliculus–Pulvinar Visual System

In all non-mammalian vertebrates the primary visual structure in the brain is called the optic tectum. In mammals the optic tectum is called the superior colliculus (SC). Another very basic visual structure in the vertebrate brain is called the dorsal posterior nucleus/anterior lateral nucleus in amphibians and fishes; nucleus rotundus in birds, reptiles, and turtles; lateral posterior nucleus or lateral posterior–pulvinar complex in non-primate mammals; and the pulvinar in primates.

The functions of the SC-pulvinar visual system are so hard to detect under normal conditions that it is often stated that this more ancient visual system has become nearly residual in primates as a result of the great expansion of the LGN visual system (e.g., Henry and Vidyasagar 1991). This is not really true, however. Both visual systems are actually larger in primates than in other mammals if we control for body size (Chalupa 1991; Garey et al. 1991; Robinson and Petersen 1992; Barton 2000).

The SC-pulvinar visual system has three major branches, one of which progresses via the retina to the SC, a nucleus in the midbrain, and then to the LGN where it becomes mingled with the LGN visual system. Another branch progresses via the retina to the SC and then to the pulvinar, another nucleus just above the LGN in the thalamus. The third branch progresses via the retina directly to the pulvinar (Figure 4.6). All branches also connect directly to extrastriate areas, such as V2 and MT, the latter

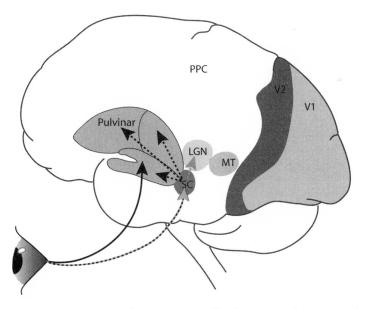

Figure 4.6   The beginnings of the superior colliculus (SC)–pulvinar visual system. Retinal connections can go directly to the pulvinar (solid black line) and SC (gray and black dashed line). The SC then provides input to the pulvinar (black dashed lines) and lateral geniculate nucleus (LGN) (gray dashed line). Not drawn to scale.

a motion-sensitive visual area (Harting et al. 1980; Perry and Cowey 1984; Jones 1985; Kaas and Huerta 1988; Garey et al. 1991; Henry and Vidyasagar 1991; Rhoades et al. 1991; Cowey et al. 1994; Williams et al. 1995; Stepniewska 2004). The SC-pulvinar visual system is the primary visual system in non-primate vertebrates (Henry and Vidyasagar 1991).

The SC itself is remarkably similar across mammals. It is separated into seven layers, often combined into superficial and deeper layers (Kaas and Huerta 1988). Cells in the superficial layers are visual in their activity, for example, responding to moving objects and changes in lighting (Kadoya et al. 1971). This visual part of the SC receives input from the eyes and sends projections to the pulvinar in addition to the LGN (Glendenning et al. 1975; Benevento and Rezak 1976; Harting et al. 1980, 1991; Huerta and Harting 1983, 1984; Lachica and Casagrande 1993; Künzle 1996; Stepniewska et al. 1999, 2000). (The superficial layers also receive input from many areas of the neocortex, including V1, V2, and, in primates, MT and PPC (Kaas and Huerta 1988), so there is also opportunity for cortical feedback to the SC.) In addition to sensitivity to motion and light, some of the well-known functions of the SC in mammals include oculomotor behaviors such as saccades, which are automatic eye movements and not under cortical control (Snyder et al. 2002). Neural connections from superficial layers to deeper layers of the SC provide a rapid route for orienting movements of the head and eyes, including express saccades, eye movements with extremely short reaction times (Isa 2002; Doubell et al. 2003).

Despite the phylogenetic conservatism of this visual system, primates have modified its functions. These changes were brought about in part by connectional modifications between the retina and the SC that appear to increase binocular input into the SC as the eyes become more convergent. The stereopsis (ability to distinguish relative depths of objects in the environment) that comes with having two convergent eyes improves our ability to break through camouflage (Julesz 1971; Frisby 1980). Stereopsis also enables faster and more accurate reaching and grasping (Melmoth and Grant 2006).

There are three important differences between the SC of primates and most non-primate mammals. First, in most non-primates, ganglion cells project from the entire visual field of each eye to the SC in the opposite (contralateral) hemisphere. In primates, however, the entire visual field of both eyes is split in half and the ganglion cells in each half project to opposite hemispheres so that a large amount of input into the contralateral SC comes from both eyes (and thus increases binocularity) (Figure 4.7). Second, in most non-primates, there is only a small projection from the ganglion cells of each retina to the SC in the same (ipsilateral) hemisphere, whereas in primates, the extent of input from the retina to the same-sided SC is much greater (Figure 4.7) (Allman 1977). Third, in non-primates the small, same-sided projection goes only to the middle portion of the SC (whereas the contralateral projection covers the whole SC). In primates the ipsilateral projection extends to the top of the SC's rostral pole (the most frontal part) where there is also contralateral input (Pettigrew et al. 1989). These changes require an evolutionary explanation.

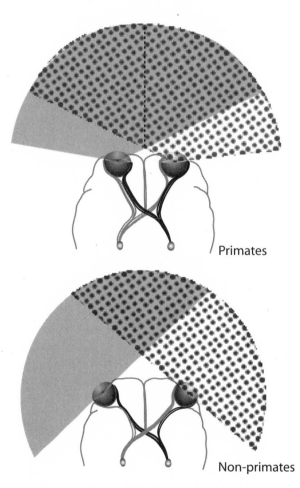

Primates

Non-primates

Figure 4.7   Retinal connections to the SC in primates compared with non-primates. The extent of retinal input into the SC of each hemisphere is balanced between the eyes in primates, whereas it is heavily biased toward the opposite (contralateral) eye in non-primates. The extent of the visual field that is shared by both eyes is much greater in primates than in non-primates.

The SC and Visually Guided Reaching and Grasping

The SC has always been considered primarily a visual structure, but our understanding of its function is beginning to change as a result of recent research. For instance, in addition to visual neurons, there are neurons that are involved in the motor behaviors of reaching and grasping (Werner 1993; Lünenburger et al. 2000; Stuphorn et al. 2000; Nagy et al. 2006). Neurons ("gaze-related neurons") have also been found in the SC that are tightly linked to both reaching and gaze (Stuphorn et al. 2000). Thus, arm movements directed to particular targets, i.e., movements similar to pointing, can speed up eye movements, or saccades, to those targets (Lünenburger et al. 2000; Snyder et al. 2002). Not surprisingly, "reach" neurons are found near "saccade" neurons in the SC (Stuphorn et al. 2000; Neggers and Bekkering 2002). We also have neurons in the SC that anchor our gaze to any target we point to (Stuphorn et al. 2000; Neggers and Bekkering 2002). Damage to the PPC, a region of the brain that is heavily involved in reaching, grasping, and pointing, can lead to our inability to point to a target to which our eyes have not fixated; with a damaged PPC, pointing to a target is possible only if the eyes have already locked onto the target (Carey et al. 1997).

Many cells in the SC's rostral pole are "fixation" neurons, which enable animals to visually lock onto a target once it has been located (Krauzlis et al. 2000). The strong binocular input to the rostral pole in primates but not in other mammals suggests that the ancestors of primates uniquely needed and benefited from a fast and accurate ability to visually lock onto targets of interest that, for other animals, would tend to blend into the background.

Many cells in the caudal pole of the SC are "movement" neurons, which fire mainly as animals make saccades to targets (Munoz and Wurtz 1995). Primates have more ipsilateral input toward the caudal pole than non-primates, and anthropoids have more ipsilateral input than prosimians, but the caudal pole is not entirely covered by input from both eyes except, perhaps, in gibbons (Pettigrew et al. 1989). Gibbons use their arms and hands to brachiate, moving hand-over-hand quickly from branch to branch in the tropical forests of Asia. Brachiation is a highly specialized form of suspensory locomotion that requires gibbons to make highly accurate reaches and grabs repeatedly to branches of varying diameters and orientations. The fast pace of their reaching and grasping may also require that their eyes move quickly to find the next target. Perhaps the rapidity with which they must search for the next handhold increased their need for greater binocular input to the caudal pole of the SC.

In fact, there is evidence that the arms and hands have influenced brain evolution not just in gibbons but also in primates as a whole. Compared with insectivores and bats, primates have expanded several areas of the brain related to motor skills together with the rest of the neocortex and thalamus (de Winter and Oxnard 2001). In addition, within the primates these structures of the brain are smallest in hindlimb-dominated leapers (such as sifakas) and scurriers (such as marmosets and tamarins), followed by runners and jumpers that use all four limbs (macaques and vervets [*Cercopithecus aethiops*]), forelimb-dominated suspensory climbers and arm-swingers (muriquis [*Brachyteles*] and gibbons), and, finally, bipedal walkers and runners (humans). This is true regardless of phylogenetic history (the degree of evolutionary relatedness). So, for example, even though the smaller-brained platyrrhine woolly monkeys and spider monkeys are only distantly related to the catarrhine chimpanzees and orangutans, the two platyrrhine monkey species converge with the catarrhine ape species in the parts of the brain that involve motor skills (de Winter and Oxnard 2001). At first glance, it might look as if the motor regions of the brain expanded more as the forelimbs became more involved in locomotion, except that this explanation does not account for the ranking of humans at the upper end since we use our arms less in locomotion than the other primates. It might be more inclusive to describe primate brain expansion as driven partly by directed forelimb action.

The PPC, the endpoint of our dorsal stream, is linked to the SC (Fries 1984; Clower et al. 2001). Thus, the SC-pulvinar visual system participates in the dorsal "vision for action" stream. The PPC, a later addition to the brains of some vertebrates, appears to inhibit the SC's automatic eye-arm-hand response (Stuphorn et al. 2000), essentially enabling us to decouple our hands from our gaze to override the functions of the more ancient SC, even to the extent that the functions of the SC are hard to detect.

The modification of the SC and the expansion of the SC-pulvinar visual system for fast, accurate, and adjustable reaching and grasping also requires an evolutionary explanation. The association between correlated brain expansion and forelimb action in locomotion, and the tight linkage between gaze and reaching and grasping, suggests that locomotion in an arboreal environment uniquely favored visually guided reaching and grasping in primates. Alternatively, because cats, which are visual predators, also have reach neurons in their SC and have visually guided reaching (Courjon et al. 2004), perhaps the first primates required visually guided reaching and grasping to capture insect prey. If this was the case, then all primates today should be able to adjust their reaching and grasping in response to moving prey. As discussed in the previous chapter, however, what little evidence we have currently suggests that primates may vary in this ability.

The SC-Pulvinar Visual System and Predator Avoidance

Old World fruit bats (family Pteropodidae) are the only mammals (other than primates) confirmed as having a primate-like SC (Pettigrew et al. 1989; Rosa and Schmid 1994). Beginning in the late 1980s there was much discussion about whether this similarity is a result of shared ancestry or evolutionary convergence (Preuss 2007). Now, however, recent molecular evidence has ruled out shared ancestry, and we must conclude that some problem that was unique to the environments of early primates and Old World fruit bats favored independent but similar solutions to that problem. While both primates and bats use their forelimbs extensively in locomotion, this probably cannot account for the convergence. Otherwise all bats, not just Old World fruit bats, would be more primate-like in their retinal projections to the SC. The similarities between the SCs of primates and Old World fruit bats are also not likely to be the result of visually guided reaching for and grasping of insect prey because Old World fruit bats are not insectivorous, they typically use their thumbs for suspension while feeding and rarely manipulate their food with their wings (Vandoros and Dumont 2004).

On the other hand, there appear to be similarities in roosting preferences that could have affected retinal connections to the SC. Early primates are thought to have been nocturnal and arboreal (see Tan et al. 2005, however, for an argument in favor of diurnal living), and so would have used trees and shrubs to roost in during the day. Similarly, most Old World fruit bats roost in trees, shrubs, and other lighted places during the daytime (with informative exceptions as we shall see in a later chapter). Roosting in lighted places during the day may expose bats to a greater risk of predation than sleeping in caves and other dark places (Baron et al. 1996), and Old World fruit bats devote a considerable amount of their daylight hours visually (and acoustically) monitoring the environment for predators (Müller et al. 2007). Vision, which would be of limited use for detecting predators in caves and other dark places, undoubtedly more effectively helps Old World fruit bats to avoid predators while roosting in the light. Presumably the same would have applied to the first primates. Thus, changes in the retinal projections to the primate SC may have initially been predator-driven, with visual guidance of reaching and grasping coming later.

Indeed, in their review of the evolution of innate predator recognition, Sewards and Sewards (2002) described a main function of the SC to be predator detection and avoidance throughout the evolutionary history of vertebrates. The deeper layers of the SC, in particular, are tuned to deal with predators. When the deeper layers are electrically stimulated in rodents, for example, the animals engage in defensive motor behavior by immediately turning, freezing, and darting. Damage to their deeper layers abolishes these behaviors to objects that loom (move fast toward them) and to objects that are on the periphery of their vision (Ellard and Goodale 1988; Northmore et al. 1988; Sewards and Sewards 2002; Brandão et al. 2003). Connections from the deeper layers of the SC go to the periaqueductal gray (PAG) and cuneiform nucleus, other parts of the brain that are also associated with most of these physical responses (Mitchell et al. 1988; Dean et al. 1989; Westby et al. 1990; Vianna and Brandão 2003). The deeper layers are also involved in what are called covert shifts of attention, when attention is directed toward salient objects before eye movements reveal shifts of overt attention (Ignashchenkova et al. 2004).

The deeper layers of the SC also have connections with the substantia nigra, a structure in the ventral midbrain. The connection from the SC to the substantia nigra allows mammals to quickly interrupt their current activity and orient to salient but unexpected objects (Comoli et al. 2003). Disinhibition of the deeper layers of the SC, to which the substantia nigra projects, causes macaques to have odd eye movements and head turning and to express fearful and defensive behavior, including exaggerated startle, cowering, and attacking of non-moving objects (Zarbalian et al. 2003). In primates the deeper layers have connections with the medial pulvinar (Stepniewska 2004), another structure of the SC-pulvinar visual system that has predator-responsive functions.

The Pulvinar and Predators

The pulvinar is located in the dorsal thalamus, and is fairly small in all mammals but primates. The entire lateral posterior–pulvinar complex in other mammals is probably comparable to just one division of the pulvinar in primates (Preuss 2007). Among primates the pulvinar is especially large in anthropoids (Walker 1938; Jones 1985; Chalupa 1991; Stepniewska 2004; but see also Chalfin et al. 2007). The pulvinar has traditionally been considered to be primarily involved in helping animals focus attention on objects that are important to them in one way or another. These objects could range from predators and conspecifics to food and substrates for locomotion. For primates such substrates would include branches of trees and shrubs.

The primate pulvinar is often divided into four different divisions: inferior, lateral, medial, and anterior (Figure 4.8). The inferior and lateral pulvinar are visual (here I distinguish between the dorsal and ventral parts of the lateral pulvinar because the dorsal part has expanded more than the ventral part in primates and they may have somewhat different functions). The medial pulvinar is multisensory (including visual and acoustic sensitivity) and the anterior is somatosensory. I will focus on the first three divisions.

One of the main functions of the inferior and ventral lateral pulvinar is to assist in selective visual processing or attention by shifting attention to relevant objects and tuning out visual information that is irrelevant (Ungerleider and Christensen 1979; LaBerge and Buchsbaum 1990; Chalupa 1991; Robinson and Petersen 1992; Robinson 1993; Morris et al. 1997; Grieve et al. 2000; Bender and Youakim 2001). In anthropoid primates it may also influence or enhance neuronal activity in V2 (Levitt et al. 1995; Soares et al. 2001a), a point to which I will return in a later chapter.

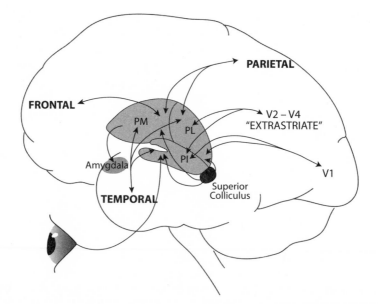

Figure 4.8   Three divisions of the pulvinar and some of their connections. The anterior pulvinar is not delineated. The medial pulvinar sends projections to the amygdala, a nucleus clearly involved in dealing with danger. PM=medial pulvinar; PL=lateral pulvinar; PI=inferior pulvinar. Not drawn to scale.

As a visual structure, the combined inferior/ventral lateral pulvinar receives input from the retina and the superficial layers of the SC. However, it also receives input from the locus coeruleus (LC) (Figure 4.8) (Stepniewska 2004). The LC will be dealt with at greater length later on but suffice it to say now that it is involved in vigilance and heightened attention to threatening situations. The primate inferior/ventral lateral pulvinar sends projections to many cortical areas, including V1 and V2 (Figure 4.8). In anthropoids those parts of V2 that receive input from the inferior/ventral lateral pulvinar show increased cytochrome oxidase metabolic activity (Ogren and Hendrickson 1977; Wong-Riley 1977; Curcio and Harting 1978; Livingstone and Hubel 1982; Wong-Riley and Carroll 1984; Levitt et al. 1995; Adams et al. 2000; Sincich and Horton 2002; Stepniewska 2004). Briefly for now, high cytochrome oxidase activity reveals regions of the brain that are more metabolically active than other regions, and implies that parts of V2 are metabolically different between anthropoids and prosimians. I will return to the relevance of this in a later chapter.

While most of the inferior/ventral lateral pulvinar appears not to have changed much over time in primates (Stepniewska and Kaas 1997; Gray et al. 1999; Soares et al. 2001b), the medial pulvinar and dorsal part of the lateral pulvinar have greatly expanded (Preuss 2007). The combined medial/dorsal lateral pulvinar is involved with many of the same functions as the SC, including eye movements and orienting to relevant objects. It is also involved in gaze direction and selective spatial attention (Robinson and Petersen 1992; Preuss 2007). Of particular interest is that of all the divisions of the pulvinar, only the medial pulvinar receives connections from the deeper layers of the SC (Stepniewska 2004), the part of the SC that is involved in defensive motor behavior. Moreover, the medial pulvinar sends projections to the amygdala, a structure also strongly involved in fearful responses (Aggleton and Saunders 2000) (Figure 4.8). If the medial pulvinar is damaged, the lateral nucleus of the amygdala, to which it projects, is not activated in the visual presence of a fearful face, and the patient does not recognize the expression as fearful (Ward et al. 2007).

Interestingly (as we will see later), the medial/dorsal lateral pulvinar does not have connections to V1 and so is not well-connected to an important part of the LGN visual system, but it does have connections with more cortical areas than the inferior/ventral lateral pulvinar (Trojanowski and Jacobson 1974; Glendenning et al. 1975; Baleydier and Mauguiere 1985, 1987; Selemon and Goldman-Rakic 1988; Garey et al. 1991; Robinson and Petersen 1992; Gutierrez et al. 2000; Stepniewska 2004) (Figure 4.8).

When spider-phobic people are shown images of spiders, brain activity increases in both the pulvinar and the amygdala (Goossens et al. 2007). Damage to the pulvinar also results in disrupted processing of threatening visual images that are presented very briefly (300 milliseconds). Without a working pulvinar, negative images that would normally draw one's attention automatically and very quickly through the SC-pulvinar visual system are not distracting until the slower LGN visual system, with its heavy reliance on cortical areas, processes the image (Ward et al. 2005).

It is intriguing that a subcortical structure involved in visual detection of danger has expanded in primates more than in other mammals and more in anthropoids than in prosimians. This suggests that something in the environment of the first primates was so novel and threatening that it favored expansion of the pulvinar to counter the threat and that later, anthropoids faced an even greater threat that favored further expansion of the pulvinar, particularly the medial/dorsal lateral pulvinar.

To summarize thus far, the SC-pulvinar visual system has always been involved in visual attention to important, novel objects that appear suddenly in the visual field (Strenge 1978; Kaas and Huerta 1988; Mollon 1989; Barton 1998; Previc 1990; Robinson and Petersen 1992). Because some of these important objects would have been predators, the SC has always been involved with predator detection and avoidance, as evidenced by its participation in evasive and defensive motor behavior. The SC was later modified in primates (and Old World fruit bats to some degree), likely in response to the greater predation risk associated with a life spent in the light, which selected for redistributed retinal connections to the SC that improved primates' ability to detect predators via greater depth perception and enhanced power to break through the camouflage of predators. The medial/dorsal lateral pulvinar also expanded in primates, most dramatically in anthropoids. It is also involved in selective attention, especially to threatening images or objects.

Thus, it should not be surprising that the SC is activated when monkeys look at snakes, which they fear (Arendes 1994), and that the SC is also activated when people see fearful faces (Vuilleumier et al. 2003). Why should we care if others are afraid? Perhaps it is because seeing fear in someone else's eyes warns us that they see something potentially dangerous nearby that could harm us. This ability to detect fear in others' faces has undoubtedly been helpful for survival and it may be mediated by so-called mirror, or empathy, neurons that are activated when we see someone perform a behavior. Our neurons fire is as if we are performing the behavior ourselves (Iacoboni et al. 2005; Iacoboni and Mazziotta 2007).

New Neural Insights on Primate Origins

Because of its noticeable expansion, the LGN visual system has long been assumed to have been of paramount importance to primate evolution. Thus, a recent study examined whether convergent orbits in primates are correlated with various aspects of the LGN visual system and other parts of the brain, including the number of neurons in the LGN P and M layers, the size of V1, the neocortex (since it is heavily devoted to vision), and larger brains overall (Barton 2004). After statistically controlling for potentially confounding factors, i.e., body size, phylogeny, and activity period (nocturnal vs. diurnal), Barton found that only the M layers of the LGN were not positively correlated with orbital convergence. This finding underscores the conservative nature of the M pathway in primates. Although the M pathway is part of the dorsal "vision for action" stream, visually guided reaching and grasping appears to be largely independent of the LGN. No "reach" neurons have been found in the LGN, for instance. Similarly, no neurons have been found in the LGN that respond when primates touch objects, as have been found in the SC. There is currently no evidence tying visually guided reaching and grasping to the great expansion of the LGN visual system.

Barton's results corroborate the evidence that the real growth of the LGN visual system has been in the P pathway. Importantly, correlated orbital convergence and P pathway expansion provide good depth perception and acute vision, and this is particularly true *for the area close to oneself and in the lower visual field* (Cartmill 1992; Barton 2004). The LGN visual system apparently did not expand for visually guided reaching and grasping for nearby food or branches; these are skills within the purview of the SC-pulvinar visual system. The evidence from Old World fruit bats suggests that the kind of rapid, accurate, automatic visually guided reaching and grasping that is required for successfully moving quickly along branches through the forest or for catching insects is a primate innovation of the SC-pulvinar visual system that was added independently of the changes that were made in the first primates for improved predator detection and avoidance. Visually guided reaching and grasping could be built onto these connectional modifications for those primates that benefited from such a skill (whether for eye-hand coordination in locomotion or for catching prey) because the ability to detect subtle differences in depth is also advantageous for quickly reaching for and grasping objects.

Thus, primates would have benefited from LGN and SC-pulvinar visual systems that were each modified to provide closer orbital convergence for the contributions that increased stereopsis provides: better depth perception and more reliable ability to break through camouflage. The greatly expanded LGN visual system wonderfully complements the more ancient SC-pulvinar visual system.

Compared with prosimians, the anthropoid LGN visual system is even more expansive and orbital convergence is even closer. We know this because the number of P layers varies within primates, and as Barton (2004) showed, the degree of orbital convergence is positively correlated with the number of neurons in the P layers in primates. This suggests that there was some greater selective pressure on anthropoids that favored even more acute vision for identifying objects and even better stereopsis for distinguishing between relative depths of objects also in the lower visual field and for cutting through camouflage of objects in the lower visual field. The presence of a fovea in anthropoids but not in all prosimians is also consistent with the idea that anthropoids uniquely benefited from clearly seeing and identifying objects that were close by and in front of them. The initial expansion of the LGN in the first primates and its further expansion in anthropoids are similar to the pattern of expansion in the pulvinar, which appears to have expanded in part to help avoid predators or other dangers.

Selective Pressures in the Lower Visual Field

What in the lower visual field could have been so important to the lives of early primates that it caused the P pathway to expand in primates? Primatologists will immediately think of ticks and other ectoparasites that are found in the hair and on the skin as a potential selective pressure. Ectoparasites negatively affect health and therefore fitness (Møller 1997), and many primates spend a lot of time grooming themselves and others to remove parasites. Grooming clearly requires focused attention, visual search, and good central vision within peripersonal space in the lower visual field. Could the benefits of parasite removal have been the ultimate cause of LGN visual system expansion? Supporting this idea is the finding that the number of grooming pairs within groups is positively correlated with group size, and a positive correlation between group size and the number of LGN P cells also exists (Barton 1996, 2000; Kudo and Dunbar 2001). Opposing this idea is the fact that greater time spent grooming in primates with larger P pathways can also be interpreted as a result of better vision, not a cause of it. Moreover, grooming is not universal among primates. For instance, grooming is virtually absent in squirrel monkeys, muriquis, woolly monkeys, and orangutans (Di Fiore and Campbell 2007; Jack 2007; Knott and Kahlenberg 2007). Most problematic, however, is that there is no a priori reason to expect parasite load to have been a greater problem for anthropoids than prosimians. Greater parasite load does not seem to be able to explain the differential expansion of the P pathway in anthropoids.

Continuing with the concept of complementarity of visual systems, an obvious alternative would be predators. Predators are consistently and universally worthy of visual attention. They are always associated with death and are thus a potentially very strong selective pressure. For that reason, most predators are best detected visually from far away. There are some, however, that are impossible, but fortunately also not necessary, to detect from far away. The danger from snakes, for example, becomes real only when they are close by. Snakes present all the conditions that a more expansive P pathway deals with effectively: they are often found in the lower visual field, they exhibit small differences in depth against their background, and they are often well camouflaged.

This is not a new idea. As part of an examination of the preferences of neurons for peripersonal distances in the P-dominated ventral visual stream of macaques, Rosenbluth and Allman (2002: 143) suggested, in fact, that near vision might be useful for seeing objects such as snakes, which "tend to creep in the lower visual field." It would clearly be helpful for primates to see snakes before stepping too close to them (or on them!). Snakes do not even have to be predators to kill. Venomous snakes, in particular, can do serious damage, even to those who are not their prey.

Here I am suggesting that the LGN visual system, like the SC-pulvinar visual system, was influenced by predator pressure. The LGN visual system would have complemented the SC-pulvinar visual system to help primates avoid being killed by predators, specifically snakes. Whereas the SC-pulvinar visual system provides a fast, automatic predator detection system with the possibility of evasive motor action, e.g., freezing or darting, the slower LGN visual system provides confirmation (or denial) of the predator and then the ability to assess the situation and act most appropriately to reduce the risk of death.

Suppression, Not Elimination

This chapter provides two examples of a more recent structure modifying the effects of a more ancient structure: the PPC can override the automatic motor responses of the SC and the LGN masks the influence of the SC, at least under normal conditions. These more powerful structures are more recent additions to the vertebrate brain. The SC (or its homolog, the optic tectum) is present in all vertebrates, including fishes and amphibians, whereas the LGN (or its homolog, the dorsal lateral optic nucleus) came into its own only after amphibians had already evolved. The PPC was added on and enlarged during the course of mammalian and primate evolution. The idea that newer structures have been laid down to override, suppress, or mask, but not replace, older structures deserves more of our attention. Until there is an object of great interest to us, or until there is a threat, the SC-pulvinar visual system seems to sit quietly in the background while the LGN visual system does its work. There are many objects in our environments that our brains detect but that we do not consciously see. Our conscious awareness of the objects around us appears to result from visual processing mainly through the LGN's P pathway, which emphasizes visual acuity, color, and object recognition via the ventral stream, when time can be taken to assess objects in the environment. The P pathway is largely responsible for the LGN visual system's great expansion in primates, and is ultimately responsible, therefore, for providing primates with the ability to be more deliberate in their assessment of objects of interest in the environment.

Allow me to speculate a bit. The LGN visual system might even be viewed as the thinking animal's vision, with greater complexity in the LGN visual system indicating greater capacity for assessment and deliberation. Perhaps the complexity of the LGN visual system can even be used as a quantitative correlate of "intelligence" between species. Intelligence is a characteristic that is hard to measure but that is evident when we are in its presence. Certainly primates are considered more intelligent than rodents or rabbits, and their intelligence is expressed partly by their curiosity, which involves scrutiny.

For our own species, vision and assessment are so interconnected that language is peppered with references to vision during thinking. In the English language, for instance, when we want to describe to someone else an idea or a concept not yet *crystallized*, we might say we are starting to *envision* or *visualize* it. When we want to let someone know we understand that scenario or concept, we often say, "I *see.*" Concepts and ideas now understood have *clarity* even though they are not objects that are visible. When we want to know what another person is thinking, we might ask them, "What is your *view* on the issue?" or "How do you *see* the issue?"

In contrast, the SC-pulvinar visual system is automatic. It does not involve assessment and does not constitute the thinking animal's vision.

In this chapter, I have tried to make very complex neural systems understandable to non-specialists. It is possible, however, that the simplified technical information I have provided is still too much for many who are not versed in basic neuroscientific jargon. If that is the case, it might help to draw attention to the two most important points in this chapter. One is that mammals, including primates, have two visual systems that do different but complementary things. The LGN visual system is more involved with becoming aware of details through vision whereas the SC-pulvinar visual system is more involved with automatic, non-conscious vision, particularly as it pertains to avoiding predators, and for primates, motor actions such as hand-eye coordination. The other key point is that primates have modified and expanded both visual systems, and the changes suggest that the ancestral environment of primates uniquely affected them to link vision with automatic, fast, accurate, and adjustable reaching and grasping, and to improve upon vision as a way to detect and avoid predators, both automatically and with conscious awareness. In the next chapter, I will provide evidence that snakes were the first, and have been the most persistent, of the modern predators of mammals, and that they indeed provided the selective pressure that favored the initial expansion of the visual sense in primates.

CHAPTER FIVE

Origins of Modern Predators

When, therefore, the earth, covered with mud from the recent
flood, became heated up by the hot ethereal rays of the sun, she
brought forth innumerable forms of life; in part she restored the ancient shapes, and in part she created creatures new and strange.
She, indeed, would have wished not so to do, but thee also she then
bore, thou huge Python, thou snake unknown before, who wast a
terror to new-created men.

Ovid, *Metamorphoses I* (~ 1 A.D.: 431–441)

T HE FIRST known fossil mammal, *Megazostrodon,* lived about
200 million years ago (Figure 5.1). It was small and probably insectivorous and egg-laying. As a vertebrate and a mammal, it
would have had both the superior colliculus (SC) and lateral geniculate
nucleus (LGN) (Butler 1994). Roughly 231–207 million years ago,
monotremes (represented today by echidnas and platypuses) diverged
from therians, the unified clade of marsupial and placental mammals
(Hugall et al. 2007 and references therein). Monotremes have a very
small LGN, and it is not clear from them that the LGN visual system initially functioned as it does today (Krubitzer 1998). Around 195–150
million years ago, therian mammals diverged into placental and marsupial mammals (Bininda-Emonds et al. 2007; Hugall et al. 2007). Because
all therian mammals minimally have the LGN, plus V1 and V2 (V1 and
V2 do not exist in reptiles), this basic mammalian LGN visual system
must have been present by at least the last common ancestor of marsupials and placental mammals (Rosa and Krubitzer 1999; Kaas 2004). If it
is true that predators were the primary selective pressure operating on
mammals to favor this initial expansion of the LGN visual system, then
we should expect some degree of coordinated timing between when
the predator appeared and when the mammalian LGN visual system
appeared, i.e., sometime before placental mammals and marsupials
diverged. If it is true that snakes, in particular, were the responsible
predators, then we should expect them to have originated around that
same time. Conversely, we should not expect the origin of the other
main predators of mammals, the carnivorans (members of the mammalian

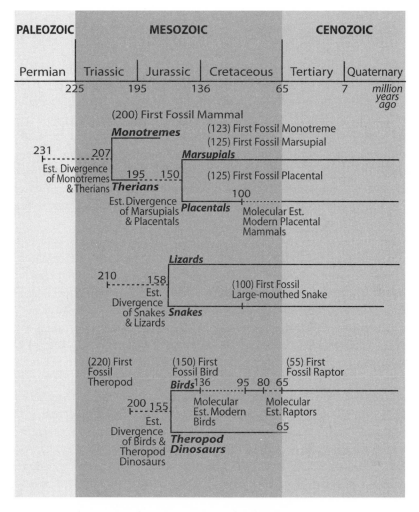

Figure 5.1 The timing of the appearance of predators relative to the divergence of marsupial and placental mammals. Molecular estimates (dashed lines) suggest that the divergence of snakes from lizards overlapped in time with both the divergence of monotremes and therians, and placental mammals and marsupials. The appearance of large-mouthed snakes corresponds more closely with molecular estimates of the appearance of modern placental mammals than do the other main predators of mammals (modern raptors and mammalian carnivores).

order Carnivora) and raptors (birds of prey), to be as closely tied to the origin of the therians.

The Origin of Modern Carnivorous Mammals

By 139–128 million years ago, early mammals had diversified to include carnivorous mammals such as opossum-sized *Repenomamus;* one specimen was found with a young dinosaur in its gut (Hu et al. 2005). Although carnivorous mammals would have been a problem for non-carnivorous prey mammals, they could not have been the predators responsible for the mammalian LGN visual system because carnivorous mammals would have already acquired the same LGN visual system from an earlier ancestor.

Crown-group placental carnivorans, the well-known lions and tigers and bears that exist today, are Laurasiatherians. Fossil evidence suggests that carnivorans evolved in Laurasia about 62–42 million years ago (Martin 1989; Van Valkenburgh 1999; Polly et al. 2006). Molecular evidence gives a broader range, about 85–55 million years ago (Wayne et al. 1989; Madsen et al. 2001; Murphy et al. 2001a,b; Bininda-Emonds et al. 2007). In either case, carnivorans originated well after the first of the crown-group placental mammals. Because modern carnivorans are crown-group mammals themselves, they simply could not have been responsible for the common LGN visual system of *all* therian mammals.

Some of the most devastating predation on primates occurs at the claws and jaws of carnivorans. These include leopards in Africa and Asia, tayras *(Eira barbara)* in South America, and fossas *(Cryptoprocta ferox)* in Madagascar. In my own field studies in East Africa, leopards decimated populations of monkeys at two of three study sites where I conducted long-term research. The leopards killed monkeys from the same group repeatedly within a span of six months or so until there were so few monkeys left that those remaining abandoned their territories and joined neighboring groups (Isbell 1994a). Not surprisingly, many primates have behavioral adaptations for dealing with carnivorans, including vocalizations that specify that the type of predator is a terrestrial mammalian predator. It would also not be surprising to find that primates (and other mammals) have visual adaptations to detect carnivorans by now, after millions of years of coexistence. Importantly, however, most such adaptations would be independently evolved modifications of the basic mammalian LGN visual system. For instance, while anthropoids, having originated in the Old World (probably Asia), would have been exposed to carnivorans for the longest period of time, lemurs would not have had to deal with them in Madagascar until 24–18 million years ago. Proto-platyrrhines would have had exposure to carnivorans in Africa but then would have been unconstrained by them for at least 23 million years after making the journey to South America because carnivorans crossed the Isthmus of Panama into South America only about three million years ago.

The Origin of Modern Raptors

Birds have been around since the time of the dinosaurs from which they evolved. In fact, they are still considered dinosaurs by many scientists. The first known bird, *Archaeopteryx,* lived about 150 million years ago (Padian and Chiappe 1998) at the late end of the divergence time frame for therians (Figure 5.1). After *Archaeopteryx,* fossils indicate that early birds diversified into waders, divers, perchers, and flightless birds. Thus far, there is no evidence that raptors had appeared by the time ancient birds began to diversify (Padian and Chiappe 1998).

There is still debate about when modern birds evolved. *Vegavis,* a fossil water bird that can be definitively called a modern bird, was recently discovered in Antarctica and has been dated to the Late Cretaceous, about 68–66 million years ago, which means that it overlapped for several million years with ancient birds whose descendants are no longer around (Clarke et al. 2005). *Vegavis* and other fossils suggest that water birds were the first of the modern birds to evolve (Feduccia 1995). Molecular studies, however, suggest that the first modern birds evolved earlier, 136–95 million years ago (Cooper and Penny 1997; Ericson et al. 2006; Hugall et al. 2007), as heavy-bodied, ground-dwelling birds (ostriches, emus, and rheas, collectively called the ratites) (van Tuinen et al. 2000). To put this in perspective relative to the appearance of the mammalian LGN visual system, if modern birds had evolved as late as 95 million years ago, they would have appeared about the same time as modern placental mammals. If they had evolved as early as 136 million years ago, they still would have appeared 14–59 million years after marsupials and placentals had already diverged and so could not have been responsible for the expansion of the LGN visual system in mammals (Figure 5.1).

Moreover, regardless of whether one feels more comfortable with fossil evidence or molecular evidence, both kinds of data support the idea that birds of prey were not the first modern birds. Raptors do not appear in the fossil record until about 55 million years ago (Feduccia 1995; Cooper and Penny 1997), while molecular estimates place the divergence of hawks and eagles from falcons in South America at 80–65 million years ago (Sibley and Ahlquist 1990; Griffiths 1999; Haring et al. 2001;

Riesing et al. 2003; Griffiths et al. 2004, 2007; Ericson et al. 2006). Owls appear to have diverged about 60 million years ago (Ericson et al. 2006). At the earliest, then, raptors could have been predators of crown-group mammals by 80 million years ago, but by then crown-group placental mammals had been around for about 20 million years with the mammalian LGN visual system well in place (Figure 5.1).

Eagles, the most dangerous avian predators of primates today, would have diverged from hawks more recently than 80–65 million years ago. Huge monkey-eating eagles live in South America (the harpy eagle [*Harpia harpyja*]), Africa (crowned and martial eagles [*Stephanoaetus coronatus* and *Polemaetus bellicosus*]), and Asia (the Philippine eagle [*Pithecophaga jefferyi*]). These eagles even prey on some of the larger primates, such as the 5–9 kilogram howler monkeys (*Alouatta* spp.), red colobus monkeys *(Procolobus badius),* and macaques. If primates have visual adaptations to deal specifically with raptors now, they did not always have them because primates had already been in existence for 10–20 million years by the time hawks and eagles evolved. Although the ability of raptors to fly makes it difficult to track their dispersal in the world and so gain a sense of the risk of raptor predation on primates on the different landmasses, it is clear that when platyrrhines arrived in South America, they had to deal immediately with predator threats from raptors because raptors evolved in South America. Today, compared with the Old World, the New World has twice as many raptor species that are documented predators of primates (Miller and Treves 2007).

The Origin of Modern Snakes

The estimated timing of the appearance of snakes fits better with the origin of mammals than the appearance of carnivorans and raptors. Molecular estimates suggest that snakes diverged from lizards 210–158 million years ago (Vidal and Hedges 2005; Wiens et al. 2006; Hugall et al. 2007), which overlaps their origin with both the time of divergence of monotremes and therians and of placental mammals and marsupials, and closer to those time periods than either raptors or carnivorans (Figure 5.1).

It is not yet clear whether snakes first evolved in aquatic environments or on land, and if they evolved on land, whether they evolved as burrowers or above-ground foragers (Greene and Cundall 2000; Caprette et al. 2004; Vidal and Hedges 2004). Wiens et al. (2006) recently presented a phylogeny of the snake-like body shape and concluded that snakes were initially burrowers, based on the basal position of extant burrowers. The most primitive-looking snake found so far also suggests a terrestrial origin. *Najash* was a legged terrestrial snake found in mid-Cretaceous rocks 99–89 million years old in Argentina (Apestiguía and Zaher 2006). It had a small mouth, as do many modern snakes that burrow and that are restricted to eating invertebrates or elongated vertebrates such as other snakes. How it killed its prey is unclear. Some extant burrowing snakes, however, use constriction to kill their prey (Greene 1983). Constriction involves winding the body around the prey or wrapping and tightening the coils around the prey, and it may have been an early innovation of snakes (Greene and Burghardt 1978). While the first small-mouthed snakes would not have been able to eat mammals, some might have been able to eat the soft and flexible eggs of smaller monotremes. *Pachyrhachis*, another legged snake, lived at the same time in marine environments. It is one of the first snakes to be found with a large gape (Zaher and Rieppel 1999). We know it could eat large prey because the gut of one fossil held a large fish (Lee et al. 1999).

The first modern-looking, legless snakes also show up in the fossil record during the mid-Cretaceous, about 100 million years ago, around the time of the initial divergence of crown-group placental mammals (Figure 5.1). Fossils of an extinct family of legless snakes called madtsoiids reveal that they arose sometime before 99 million years ago and lived until about 55 million years ago, mainly in Gondwana. They have also been found in southern Spain and Romania (Rage et al. 2004; Folie and Codrea 2005), and probably arrived in Europe from Africa (Rage 1996; Scanlon and Lee 2000; Rage et al. 2004). Madtsoiids were small to very large (from half a meter to nine meters, equivalent to 1.5 feet to nearly 30 feet) aquatic or terrestrial snakes that did not burrow. They had a large gape and no legs and are thought to have been constrictors (Vidal and David 2004; Scanlon 2006). Madtsoiids are sometimes viewed as having been similar to boas and pythons, the constrictors that exist today and that are confirmed predators of mammals, including primates. If the fossil record shows constrictors as present at about 99 million years ago, chances are that they evolved earlier.

Based on their distribution and genetic relationships, it is likely that modern constrictors evolved in Gondwana (Noonan and Chippindale 2006). Boas are largely found in South America, a Gondwanan continent, and the constrictors that live in Madagascar, another Gondwanan landmass, are closely related to South American boas (Kluge 1991; Vences et al. 2001; Noonan and Chippindale 2006). Pythons may have originated in Africa, yet another Gondwanan continent, or Asia, before dispersing to Australia (Rawlings et al. 2008). Fossils of modern constrictors have been found in Laurasia, but may be no more than about

65 million years old (Rodriguez-Robles et al. 1999). If one agrees with the origin of crown-group placental mammals in Gondwana, then the first crown-group placental mammals were present with constrictors in Gondwana at about the same time. Indeed, fossils of snakes and placental mammals have been found together in at least one Cretaceous fossil site in India (Prasad et al. 1994; Khajuria and Prasad 1998; Rage et al. 2004), a confirmation that snakes and mammals did encounter each other very early on.

Colubroid snakes are the other mammal-relevant extant group of snakes. Colubroids are the so-called advanced snakes, e.g., vipers, cobras, and other venomous snakes, as well as racers, gopher snakes, and other non-venomous snakes. The earliest advanced snakes originated in Asia (or possibly Africa) and were probably nocturnal (Cadle 1988; Vidal and Hedges 2002; Vidal et al. 2007). Fossils of putative colubroids have been dated to the mid-Cretaceous, 99–93 million years ago, in northern Africa (Rage and Werner 1999; Rage et al. 2003). In contrast to boas and pythons, which always constrict their prey, some colubroids do not constrict but merely bite to subdue (Greene 1997).

By about 100 million years ago, snakes had diversified into a variety of forms: small- and large-gaped, aquatic and terrestrial, legged and legless. Whether snakes initially evolved on land or in water, it is indisputable that *when the first of the modern placental mammals evolved, they lived in an environment where snakes looked as they do now.* All had elongate shapes and all undoubtedly had scales because snakes diverged from lizards, which also have scales. This was at least 20 million years before modern raptors appeared and 15–45 million years before modern carnivorans appeared. If you prefer to think that crown-group mammals evolved as early as molecular studies suggest but in Laurasia, then crown-group placental mammals would have been joined by early constrictors and perhaps early colubroids. Alternatively, if you prefer to put all your faith in fossils and believe that crown-group placental mammals evolved in Laurasia about 65 million years ago, they still would have had to deal with snakes before raptors or carnivorans appeared. The point to remember here is that whether crown-group placental mammals evolved 100 or 65 million years ago, in Gondwana or Laurasia, snakes that could have eaten them were present before the other predators appeared.

Most herpetologists think that snakes evolved large gapes to take advantage of mammals as a new food source and that the later diversification of snakes in the Paleocene was a response to the concomitant diversification of mammals (Greene and Burghardt 1978; Greene 1983; Rodriguez-Robles et al. 1999; Rage and Escuillié 2000; Lee and Scanlon

2002; Scanlon 2003; Vidal and David 2004). When a new prey type appears, natural selection may modify predators in response, resulting in new adaptations to catch and eat the new food type. Of course, natural selection also operates on prey animals, with the result that they may also evolve ways to minimize predation. Therein lies what is known in ecological circles as the predator-prey co-evolutionary arms race. Modification of visual systems may be one expression of the predator-prey relationship. For instance, in response to predators, better vision might evolve in the prey to make it easier for them to detect predators, and then predators might counter with adaptations to make themselves more difficult to see.

Snakes are indeed remarkably good at keeping themselves hidden from view. They are nearly impossible for mammals, including primates with their excellent vision, to see from a distance. Fortunately this is not a terrible problem for their prey because snakes are dangerous only when they are within fairly close range. On the other hand, snakes can be highly cryptic even up close. The ability of many snakes to blend into the background depends on both their coloration and the disruption to their patterns of color as they lie coiled upon themselves (Coss 2003). Snakes often have color patterns that camouflage themselves very well against the background substrate.

Camouflage works best against those with good color vision (more on this later in the book), and animals with excellent color vision include snakes and other reptiles, birds, and many primates (but no other mammals). Camouflage can be useful against both predators and prey because in both cases there is an advantage in not being detected. Snakes eat birds and other reptiles in addition to primates. However, birds, other reptiles, and primates (although primates less so) also eat snakes. Thus, camouflage in snakes could have evolved because it helped them avoid detection both by those that would prey on them and by their own reptilian and avian prey. Importantly, camouflage probably did not evolve in snakes to avoid detection by their non-primate mammalian prey, which do not have good color vision and so should be better able to cut through camouflage. Among those mammals that successfully avoided being eaten by snakes were those that could visually detect snakes quickly and that were sensitive to the visual cues that snakes always provide, independent of coloration. The most important point here is that the visual cues that snakes provided mammals from the beginning are cues that snakes still have today.

Visual cues that could be attended to by both visual systems would have been ideal, with the automatic SC-pulvinar visual system offering mammals a slightly shorter reaction time than they would have if they had to rely entirely on the LGN visual system, and the LGN visual system offering mammals the time to assess the situation once the snake has been perceived. If you do not think the difference in processing speed between these two visual systems is long enough to matter, ask yourself if you would take less than a half-second lead if you were about to step on a black mamba *(Dendroaspis polylepis)* or a bushmaster *(Lachesis muta)*.

CHAPTER SIX

Vision and Fear

But our reason telling us that there is no danger does not suffice. I may mention a trifling fact, illustrating this point, and which at the time amused me. I put my face close to the thick glass plate in front of a puff-adder in the Zoological Gardens, with the firm determination of not starting back if the snake struck at me; but, as soon as the blow was struck, my resolution went for nothing, and I jumped a yard or two backwards with astonishing rapidity. My will and reason were powerless against the imagination of a danger which had never been experienced.

Charles Darwin, *The Expression of the Emotions in Man and Animals* (1872: 38)

MANY CONSERVATIVE features about visual systems in mammals may reflect shared evolutionary history. For example, in the marsupial and placental mammals that have been studied, cells in both the lateral geniculate nucleus (LGN) and V1 are responsive to lines of different orientations (Marrocco 1972; Norton and Casagrande 1982; Ibbotson and Mark 2003; Kaas 2004; Heimel et al. 2005; Van Hooser and Nelson 2006). These types of cells are likely to have existed in the last common therian ancestor (Ibbotson and Mark 2003). Visual systems are also modifiable, however, depending on the selective pressures operating in particular environments. The variation in orbital convergence that we see in mammals is one example, the near loss of eyesight in subterranean mammals another (Cooper et al. 1993; Crish et al. 2006).

If it is true that snakes, as the first and most persistently present of all the predators of mammals, were responsible for the initial expansion of the LGN visual system in the last common ancestor of mammals living today, then we should expect all mammals today (except perhaps those with degenerated visual systems, e.g., naked mole rats [*Heterocephalus glaber*]) to have visual neurons that are stimulated by the cues provided by snakes.

For example, the sensitivity to lines of different orientation should be beneficial for detecting the scales on snakeskins.

The superior colliculus (SC), pulvinar, LGN, and visual areas V1 and V2 exist in all therian mammals (Rosa and Krubitzer 1999). If snakes influenced visual systems from before the divergence of marsupials and placentals, and if snakes are still operating today as a selective pressure on mammalian visual systems, cells in therian visual structures (if not all mammals) should respond to visual cues given by snakes regardless of the mammal species examined. Highly periodic patterns are predicted to be one such cue; these are common in snakes but otherwise infrequent in nature (Coss 2003). At the most basic level, highly periodic patterns of snakes incorporate lines of different orientations, edges, contours, and corners. All are seen in the scales of snakeskins, and all have been found to stimulate neurons in the early parts of mammalian visual systems. These classic response properties of neurons are not usually considered for their relevance to natural stimuli, and especially not to snakes, but there can be no doubt that there must have been selection on visual systems for the detection of what is salient to animals. The fact that different taxa of mammals still have such neurons after many millions of years of divergence suggests that either selection cannot get rid of those neurons or there is still a selective advantage in having them. Given that other, more complex traits are often successfully reduced or lost when they are no longer selectively advantageous (e.g., the reduced eyes of naked mole rats), it seems worth considering that a selective pressure indeed continues to maintain cells such as orientation-sensitive cells in all mammals, in all their environmental niches, despite all these many millions of years. As a selective pressure on mammals, snakes have been one clear constant from the time of the therian divergence to the present.

All therian mammals have similar, basic neural structures for detecting and avoiding animals that are dangerous to their survival. Obviously, those individuals that did not respond to potential predators by detecting and avoiding them in the past would not have survived long. Conversely, those that were able to detect and avoid predators would have had a greater chance to survive long enough to reproduce and pass on whatever genetic material they had that enabled them to be more sensitive to predators. One way to avoid predators (although of course not the only way) is to see them first, and so, all else being equal, visual systems of animals should have become modified over evolutionary time to speed up predator detection and make identification more reliable. For some of us, all snakes need to do to inspire fear or other responses is to be within our sight. Indeed, we mammals may be highly sensitive to visual cues of snakes. For example, California ground squirrels *(Spermophilus beecheyi)* and wood rats *(Neotoma albigula)* react fearfully to snakes on the first day that they can see (Coss 1991, 2003). The SC of a squirrel, by the way, is up to 10 times larger than that of a rat of the same body size (Lane et al. 1971; J. Kaas, pers. comm., 21 March, 2008). Being diurnally active, squirrels were probably able to take advantage of the light (as did Old World fruit bats that roost in the light) and expand their SC-pulvinar visual system under heavy evolutionary pressure from their predators.

In addition to providing simple visual cues of the components that comprise their skins, snakes provide cues that could be used to gain a gestalt of whole snake. Thus, cells in the SC, pulvinar, and V2 (the V2 receiving complementary input from the SC-pulvinar and LGN visual systems; Sinich and Horton 2002) are capable of integrating spatial and form visual cues (Benevento and Port 1995) that reflect *global* rather than *local* features (Merigan et al. 1993; Peterhans 1997; Roe and Ts'o 1997; Kaas 2004) and that should help mammals identify snakes quickly. In cats, for example, some cells in the SC are sensitive to coherently moving short lines but become more sensitive to individual moving lines when they are longer (Zhao et al. 2005). In the context of snakes, the former should be useful for identifying snakes by their scales (which are shorter and always move together) and the latter by their bodies. In cats and humans pulvinar cells respond preferentially to moving or flickering plaid patterns (Casanova et al. 2001; Kastner et al. 2004; Villeneuve et al. 2005), which are strikingly reminiscent of the tessellated image produced by scale patterns on moving snakes (Figure 6.1). Most of the plaid-sensitive neurons are found in the medial pulvinar (Casanova et al. 2001; Kastner et al. 2004), the division that connects with the amygdala, a structure involved with fearful behavior. Visual sensitivity to the cues given by snakes

Figure 6.1 Plaid and checkerboard patterns that elicit a response from select pulvinar, V4, and inferotemporal (IT) cells compared with a snakeskin. The similarities are striking.

could explain the innate recognition of the snakeskin pattern in California ground squirrels and wood rats (Coss 1991, 2003).

The visual systems of primates have been more extensively studied than those of all other mammals. In macaques, V2 cells respond preferentially to short lines, corners, contours, occlusion (which requires good depth perception), and movement of elongated objects (Peterhans 1997). These sensitivities should be helpful for detecting the edges and angles of snake scales, the elongate shape of snakes, their movements, and their bodies when they are partially hidden by grass or other vegetation. Cells in the V2 also respond preferentially to small spots of color within larger receptive fields (called spot cells) (Roe and Ts'o 1997). As part of their camouflage, snakes often have on their skins patterns of small spots of color set against a contrasted background. Finally, V2 cells respond preferentially to rows of spots moving together against a background (called coherent motion cells) (Peterhans and von der Heydt 1993; Peterhans 1997). This image is described as an oscillating rod-like object (Peterhans 1997), a description that might form the image of a moving snake in one's mind. Neurons in V2 are also more binocularly responsive than V1 neurons, and allow greater separation between the object and its background (Peterhans and von der Heydt 1993). This provides for good depth perception and should be useful for distinguishing snakes from their substrate, a difference often involving only a couple centimeters.

Primates have more cortical areas devoted to vision than other mammals, and some of these areas may also be responsive to attributes of snakes. For instance, some cells in V4 and inferotemporal cortex (IT) respond preferentially to patterns that resemble larger segments of geometric patterns on snakeskins (Figure 6.1). In V4 neuronal activity increases more for checkerboard patterns (highly periodic patterns) than uniform textures when both are shown in the peripheral visual field (Kastner et al. 2000), a sensitivity that might enable us to detect snakes out of the corners of our eyes. In the IT a diamond shape evokes a greater neuronal response than a circle, a triangle, or random dots (Okusa et al. 2000). Snake scales are often diamond-shaped. Cells in the IT also still respond to patterns composed of simple oriented segments (another highly periodic pattern) when they are shown so quickly to the observer that there is no conscious awareness (Kovács et al. 1995a). Finally, cells in the IT also respond to shapes that are partially hidden (Kovács et al. 1995b), recalling conditions surrounding the proverbial snake in the grass.

These tantalizing tidbits begin to move visual neuroscience toward the Darwinian paradigm and the reality of the natural world. There had to be some environmental stimulus that helped form the preferred responses of visual cells, but an evolutionary perspective is still rare among neuroscientists. Here I am suggesting that snakes helped direct the evolution of some visual cells to become sensitive to their physical characteristics and that the cells' sensitivity to those physical cues primes mammals for quick snake detection and identification. Of course, this is not expected to happen to the exclusion of being able to detect and perceive other objects. Obviously, other natural objects and scenes also have lines, edges, and contours. But how many have diamond-shaped patterns and elongate shapes? Taken together, the response properties of many cells in both SC-pulvinar and LGN visual systems describe the coherence of visual attributes of snakes as well or better than other objects that might be universally important to all mammals, and better than food and branches, the objects that have been claimed as vitally important in hypotheses about primate origins.

Neuroscientists have not suggested that cell response properties are related to characteristics of snakes nor have they tested cells with that idea in mind. We are probably just lucky that some of the visual stimuli that have been used in neuroscientific research are images that simulate snake characteristics. Or, maybe it was not just luck. Perhaps there is more to it than we realize. Consider the possibility that we are drawn to those stimuli because they are visual signals that automatically attract attention.

The Vertebrate Fear Module

I would like to focus the remainder of this chapter on automatic visual detection as one way that snakes capture attention. The SC-pulvinar visual system is the visual part of the fear module, a behavioral and neural system that is automatically activated (Öhman and Mineka 2003). In other words, an animal does not need to have a neocortex to visually detect and respond appropriately to threats. This makes sense because nonmammalian vertebrates do not have a neocortex and yet they are still able to respond appropriately to their predators. In addition to the SC and the pulvinar, the fear module includes the amygdala and the locus coeruleus (LC). The Appendix encapsulates the information in the following sections and Figure 6.2 provides a visual image of where some of these structures are located in the brain of a macaque and of some of the neural connections of these structures. Despite the importance of the LGN visual system in predator identification, it is not part of the fear module. Although the LGN visual system may also deal with fear, it is not a part of the fear module because it is involved in conscious vision and because it can override the fear module.

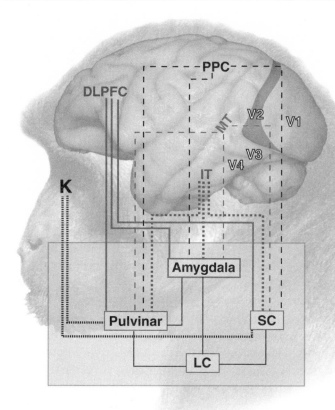

Figure 6.2   Representative primate brain showing the relative locations of the various structures of the fear module (and associated cortical areas) and its connection to vision, particularly the K pathway. See the text for discussion of the K pathway. SC = superior colliculus; LC = locus coeruleus; V1–V4 = visual areas 1–4; IT = inferotemporal cortex; MT = middle temporal cortex; PPC = posterior parietal cortex; DLPFC = dorsolateral prefrontal cortex. Connecting lines do not always imply directionality.

The Amygdala

The amygdaloid complex exists in all vertebrates. In mammals, it is located in the temporal lobe and has connections to many parts of the neocortex (Amaral and Price 1984; McDonald 1998). The mammalian amygdala is composed of several nuclei. The central nucleus mediates motor expressions of fear such as sudden inhibition of movement, i.e., freezing (Kalin et al. 2004). It has connections with the basolateral nucleus of the amygdala as well as the LC, substantia nigra, and periaqueductal gray (PAG), other structures also associated with the fear response (Jones and Burton 1976; Iwai and Yukie 1987; Amaral et al. 1992; Aggleton and Saunders 2000; Davis 2000; Kalin et al. 2004). The central nucleus appears to be conservative among mammals, being no larger in primates than in insectivores once differences in body size are considered (Barton and Aggleton 2000).

The lateral nucleus is involved with both hearing and vision, and in primates is particularly involved with that aspect of vision involving gaze direction (Brothers et al. 1990; LeDoux 2000). The lateral nucleus is the only nucleus of the amygdala with direct connections to the medial pulvinar. In addition, the lateral nucleus receives input from the LC and from cortical areas. It sends projections to the amygdala's central and basolateral nuclei (Jones and Burton 1976; Amaral et al. 1992; Aggleton and Saunders 2000; LeDoux 2000; Pitkänen 2000).

The basolateral nucleus (also known as the basal nucleus) helps animals learn about threatening stimuli, acquire and express fearful behavior, acquire spatial memory, and modulate memory storage, especially of emotionally charged stimuli (Cahill and McGaugh 1998; LeDoux 2000; McGaugh et al. 2000; Berridge and Waterhouse 2003; Zarbalian et al. 2003). The basolateral nucleus is enlarged in tarsiers and anthropoids but not in non-tarsier prosimians (Barton and Aggleton 2000; Barton et al. 2003). The enlargement of the basolateral nucleus is revealing. It would not likely have expanded in size had primates not had to deal with a new selective pressure. In primates the basolateral nucleus receives input from the LC, posterior parietal cortex (PPC), and the dorsolateral prefontal cortex (DLPFC), and sends projections to many areas, including the V1, middle temporal area (MT), and, within the amygdala, the central nucleus (Selemon and Goldman-Rakic 1988; Aggleton and Saunders 2000).

Overall, it is clear that the amygdala is important for learning what is threatening and for responding appropriately (Jones and Burton 1976; Morris et al. 1997, 1998; Cahill and McGaugh 1998; LeDoux 2000; Kalin et al. 2001; Amaral 2002). The amygdala plays a major role in helping animals survey and evaluate the environment for danger signals (Amaral 2003), a task that combines spatial and perceptual visual abilities.

The amygdala also enables primates to use cues provided by others to help them detect fear-related stimuli preconsciously (Morris et al. 1999; Dolan and Vuilleumier 2003). While the jury is still out on prosimians, anthropoid primates are clearly able to orient their eyes in the direction of the gaze of others, using the head and eyes as cues (Kawashima et al. 1999; Emery 2000; Scerif et al. 2004; see also Vuilleumier 2002). A change in the direction of another person's gaze triggers an automatic redirection in our own gaze (Driver et al. 1999; Langton et al. 2000). For us, eyes alone can be strong cues for detecting danger in the environment; we need only see fearful eyes to get a reaction from our amygdalas (Morris et al. 2002). People with damaged amygdalas are able to detect happiness in faces but not fear (Amaral 2003). Although the eyes alone are able to convey fear, people with this deficit miss the information because they do not automatically focus on the eyes, suggesting a lack of SC-pulvinar involvement. By definition, an automatic response occurs before or without conscious awareness (Dolan and Vuilleumier 2003). Once people with amygdaloid lesions are made aware of the eyes by being told explicitly to look at them (thereby overriding the automaticity of the SC-pulvinar visual system), they can recognize fear (Adolphs et al. 2005).

The Locus Coeruleus

The LC is a structure near the cerebellum, the neuronal activity of which is driven by salient and threatening or alarming objects (Berridge and Waterhouse 2003). In rats, lesions of the LC result in reduced freezing behavior toward threatening objects (Neophytou et al. 2001). The LC is the primary source of norepinephrine in the mammalian brain, a neurotransmitter that is associated with increased attention or vigilance, increased learning, and enhanced memory, particularly of negative experiences (Aston-Jones et al. 1991, 1994, 1997; Foote et al. 1991). Without norepinephrine in the amygdala, we cannot express fear (Schulz et al. 2002). The LC has widespread connections with other structures of the fear module, including the SC, pulvinar, and amygdala, and with early parts of the LGN visual system, including the V2. The LC has strong connections with the LGN itself, but strangely, only in non-anthropoid mammals (Morrison and Foote 1986; Wilson et al. 1995; Berridge and Waterhouse 2003).

Other Brain Structures Involved in Fear

It is important to realize that other structures besides the SC, pulvinar, amygdala, and LC influence behavioral responses to predators and other threats. These other structures include the PAG, hippocampus, hypothalamus, cingulate cortex, intralaminar thalamic nuclei, and the parabigeminal nucleus, all of which have connections to one or more of the basic structures of the fear module. Also, I have only briefly mentioned some of the connections among the basic structures of the fear module and other parts of the brain. In fact, nearly all the basic structures have connections to many cortical areas in mammals that undoubtedly help to coordinate appropriate responses to predators.

In primates, for example, some cortical areas, such as the V2, MT, and the frontal eye fields, may assist in the initial or global perception of predators. Other areas, such as the V1, V4, prefrontal cortex, and anterior cingulate cortex, may be involved in refining that perception and processing information once attention has been drawn to the threat (see Shulman et al. 2001). Perhaps this is where evaluation and consideration of alternative responses take place. Other areas, such as the PPC, are involved in spatial perception, spatial attention, and spatial memory, and may help primates to remember and avoid places where they have recently seen predators. It should be fairly easy to remember and avoid the locations of snakes because snakes move relatively little and have small home ranges compared with most primates.

One of the ways that animals can detect danger in their environment is by monitoring the behavior or expressions of others. Parts of the fear module (the pulvinar and the lateral amygdala) and some of the cortical areas to which they are connected (e.g., PPC) are involved in gaze direction. Snake-naïve macaques develop fearful responses to snakes when they watch other macaques on videotape reacting fearfully toward snakes; they do not develop fearful responses to flowers when they watch spliced videotapes of others expressing fear toward flowers (Cook and Mineka 1989). This difference shows that context is also important, at least for monkeys. Flowers will not, and never have, hurt monkeys, but snakes have. Monkeys are primed to be afraid of snakes, and they can often learn to be afraid by cueing in on, or gazing at, others.

Many areas of the mammalian brain have been modified and expanded in primates, and many other areas of the brain in primates have no counterpart in other mammals, including a number of visual areas (Preuss 2007). The observation that naïve macaques learn to react fearfully to snakes after observing the fearful reactions of others tells us that there is a certain amount of learning and cortical involvement in primates' behavioral reactions to snakes (Cook and Mineka 1989). Cortical involvement is further illustrated by the finding that glucose metabolism, a sign of brain activity, increases in the DLPFC when macaques are exposed to snakes but not when they are exposed to non-threatening objects, or when amygdala-damaged macaques are exposed to snakes (Roberts et al. 2002).

Obviously, this is not to say that the visual systems and associated cortical areas were not also useful for or modified to accommodate other purposes over time. The PPC provides an irrefutable example of a brain structure initially used for one purpose that later evolved to become useful for another related purpose. In non-human primates, recall that the PPC is activated during reaching and grasping with the arms and hands. In humans, however, it is also activated when people name objects, especially tools that they use with their hands (Chao and Martin 2000). And while I mentioned earlier that the PPC is involved in spatial perception, spatial attention, and spatial memory, abilities that should be useful to primates for remembering and avoiding the locations of predators, they should also be useful for traveling in the complicated three-dimensional environments of tropical forests and for remembering the locations of patchily distributed foods. Similarly, although the SC-pulvinar visual system may have enabled primates to respond automatically to the gaze of others (the mother, initially) to look for snakes, and the LGN visual system may have enabled them to identify snakes, they might have later responded to the gaze of others in threatening social situations (e.g., Emery 2000), particularly when they began to live in groups. Group living would have added a sociopolitical context to visual systems that could then have led to even further brain expansion in primates (Dunbar 1998).

The K Pathway

How does the retina pass visual information to the fear module so that animals detect and respond automatically to the threat? There must be at least one visual pathway from the eyes to the SC and pulvinar, the visual parts of the fear module. This visual pathway may be the koniocellular (K) pathway (W in non-primates) because it provides the main retinal connections to the SC and pulvinar (Figure 6.3).

A word of caution: I am labeling this visual pathway the K pathway but this may not be appreciated by neuroscientists. In essence, I am combining several projections from the retina into one pathway largely based on the fact that certain retinal ganglion cells (i.e., gamma, W-like, or K cells) go to the SC, pulvinar, and K layers in the LGN. For simplicity's sake I am calling these ganglion cells "K retinal ganglion cells" or "K cells." In the future, this pathway may end up being split into several pathways as we learn more about the functions of the K cells; as we know it today, the K pathway is certainly heterogeneous in its behavior (Rodman et al. 2001; Casagrande et al. 2007).

The K pathway has three routes. The first route goes from the retina to the SC. In fact, the SC gets most of its visual input from K retinal ganglion cells. A minority are M cells, which are likely involved in aspects of the automatic, visually guided reaching and grasping described earlier. There may also be a sprinkling of P cells (Williams et al. 1995). The K pathway is the only visual pathway known to connect the SC with the LGN (Casagrande 1994; Preuss 2007), enabling the SC to communicate directly with the LGNs K layers.

A second route of the K pathway goes to the pulvinar. Like the SC, the pulvinar receives most of its retinal input from K ganglion cells (Cowey et al. 1994; Stepniewska 2004). Given that most of the visual input to the SC and the pulvinar comes from K cells, it stands to reason that the K pathway is more involved in the SC-pulvinar visual system and, thus, the fear module, than the M and P pathways.

The third route of the K pathway goes from the retina to layers in the LGN called K or S layers (in keeping with the naming system for the other visual pathways, the K pathway is named for its projections from the ret-ina to the LGN). Sometimes the layers are not recognized as such and the K retinal cells are said to project to interlaminar cells (Hendry and Reid 2000). These layers or interlaminar cells separate individual M and P lay-ers from other M and P layers (see Figure 4.3). Thus, as the P pathway has

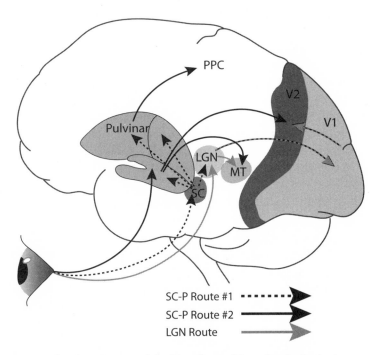

Figure 6.3   The three routes of the K pathway. Note the similarity in connections of the K pathway and the SC-pulvinar visual system in Figure 4.6.

expanded in primates, so has the K pathway. Because catarrhines have the greatest number of P layers (up to six), they also have the greatest number of K layers (Hendry and Reid 2000). From the LGN, some neurons in the K pathway project directly to the MT (Sincich et al. 2004), an area that is heavily involved in motion. The K pathway's counterpart in non-primates, the W pathway, mainly goes from the retina to the SC, with a minor projection to the LGN (Henry and Vidyasagar 1991).

Until recently the K pathway was thought to be almost non-existent in primates (Henry and Vidyasagar 1991; Hendry and Reid 2000), but this was partly a result of greater research focus on the P and M pathways and also partly a result of the difficulty of examining the very small cells that are involved in the K pathway (Casagrande 1994; Kaplan 2004). Recent research, however, has begun to reveal that the K pathway is not only present in primates but also quite complex.

The K cells are more heterogeneous than M and P cells. Their known functions include sensitivity to differences in luminance and very rapid responsiveness and visual processing of motion via direct projections from the LGN to the MT. The K cells can thus be viewed as overlapping somewhat with the function of the M pathway. Koniocellular neurons also respond to hues of blue and even sounds (Irvin et al. 1986; Casagrande 1994; Morand et al. 2000; Calkins 2001; Rodman et al. 2001; Casagrande and Royal 2003; Sincich et al. 2004). The K cells involved with the color blue seem to be restricted to the retinogeniculate route (Hendry and Reid 2000). The K pathway may also be involved in object recognition via its projections from V1 to extrastriate areas in the temporal cortex (Casagrande 1994; Casagrande and Xu 2003). Object recognition is a classic function of the P pathway. Finally, the K pathway is likely to be involved with saccades and attention, aspects for which the SC is well known (Allman and McGuinness 1988; Mollon 1989; Lachica and Casagrande 1992; Casagrande 1994, 1999; Martin et al. 1997; Reid et al. 1997; Hendry and Reid 2000; Shostak et al. 2002; Casagrande and Xu 2003; Chatterjee and Callaway 2003).

People whose SCs have been damaged are not able to detect stimuli that normally pop out from the background, such as targets embedded in distractors (say, an L within a field of Ts) (Figure 6.4) or brief flashes of light in their peripheral vision. Damage to the SC likely involves disruption of the K pathway. When the LGN visual system's P pathway is disrupted, it also results in severe visual deficits. This will not be surprising when you recall that the P pathway is responsible for our visual acuity. What is surprising is that damage to the P pathway does not necessarily result in total blindness, indicating that there is a kind of vision that re-

Figure 6.4   Example of pop out. People whose SCs are damaged have a more difficult time detecting lines and letters that are different from the rest. For normal people, the one L in this figure quickly pops out against a background of Ts.

veals itself when the P pathway, our most dominant visual pathway, is quiescent and not overwhelming it (Rafal et al. 1990; Sincich et al. 2004). This is shown by studies of blindsight.

Blindsight and the K Pathway

In humans damage to the V1, to which both P and M (as well as K) cells project from the LGN, results in a loss of conscious visual awareness, or what has become known as blindsight. Importantly, when V1 lesions occur they result in the degeneration of most of the P layers in the LGN and most of the P ganglion cells in the retina (Stoerig and Cowey 1993; Cowey et al. 1994). People with blindsight are often still able to locate objects despite having no conscious awareness of having seen them (Weiskrantz et al. 1974, 1995; Barbur et al. 1980, 1999; Blythe et al. 1987; Stoerig et al. 1997). So, for example, if you were to hold up a pencil and move it past such a person's blind visual field, that person might say that she did not see anything, but she might also say she had a feeling or a sense that something had moved in front of her. Blindsighted people often make saccades to the location of the target even though they insist that they cannot see anything. Some people with blindsight can also point to the object they cannot see. Recall that the SC-pulvinar visual system is very much involved in saccades and gaze-connected pointing. Blindsight also occurs in macaques (Cowey and Stoerig 1997). Macaques with damaged V1s can still move their heads away from things that suddenly expand, or loom, at them (King and Cowey 1992).

Brain-imaging studies have confirmed the involvement of the SC-pulvinar visual system in blindsight and masked conditions that mimic blindsight (Sahraie et al. 1997; Liddell et al. 2005; de Gelder and Hadjikhani 2006). Because the majority of retinal ganglion cells that project to the SC and pulvinar are K cells, it seems reasonable to suggest that the K pathway is involved in blindsight (although perhaps not to the total exclusion of the other pathways). The K pathway's involvement in blindsight may be possible either because retinal K cells project to the SC and pulvinar, which then project directly to extrastriate areas beyond the V1, or because LGN K cells, which also receive input from the SC, also send projections directly to the extrastriate cortex, such as the MT (Yukie and Iwai 1981; Bullier and Kennedy 1983; Cowey and Stoerig 1989; Stoerig and Cowey 1993; Bullier et al. 1994; Rodman et al. 2001; Sincich et al. 2004).

The K Pathway and Preconscious Detection of Snakes

One evening I was talking to Allen Herre, a biologist at the Smithsonian Institution's Tropical Research Institute on Barro Colorado Island (BCI), Panama, about my new hypothesis linking snakes and the evolution of primate visual systems. I was not doing a very good job of getting him interested. He was tired, or bored, or both. But when I got to the part about our ability to detect snakes before we are actually aware of them, he sat bolt upright from his supine position on the couch and excitedly described what he called one of the most extraordinary experiences of his life. He had just finished a long hard day of work on BCI and it was getting dark. He started to run through the forest toward the trails by way of a vine tangle. He had gone a little way when he suddenly froze:

> The switch over from wild abandon "we are finally done" no cares running through the forest to . . . stopping dead in my tracks and having every sensory system being on absolute high alert without having any idea what it was on alert for . . . no I will not forget that.

The green vine snake was at eye height right in front of him when he finally saw it.

When snake-phobic people are shown pictures of both snakes and mushrooms using backward masking, a technique that mimics blindsight by presenting the image of the object so briefly that people are not aware that they have seen the images, we know that their brains have detected the snakes because they respond with physiological indicators of anxiety only to the snakes (Öhman and Soares 1993, 1994). Because V1 is inactivated by blindsight, and blindsight involves detection but not awareness, it may be that V1, the largest cortical visual area in mammals, is important for conscious *awareness* of snakes, for example, but not for their preconscious *detection*. It may be that the preconscious detection of snakes in particular requires stimulation of cells that respond to diamond shapes or elongate objects, for instance, types of cells that are very rare or absent in V1. In contrast, cells in the V1 may be best for the kind of vision that involves scrutiny, the task of focusing on details in the visual field. Once the possibility of a snake is detected by the K pathway via the SC-pulvinar visual system, V1 could ascertain whether it is indeed a snake by processing finer details, as the following quote suggests.

> [The banquet] lasted, I remember, well beyond midnight, and had become musical and maudlin when the oil lamps, in their zareba of fluttering insects, began to show signs of exhaustion. As they flickered and dulled, there suddenly came the cry "Cobra!" There was pandemonium at once, and just before the lights went out altogether the shape of the reptile was seen in a corner. Some climbed on chairs, others on tables. A few bold spirits grasped sticks and fiercely attacked the serpent. It leapt and twisted beneath their blows, and then everything was plunged into darkness.
>
> Shouts of advice came from outside, where half the town had gathered. From inside came repeated cries for lights—more lights—quick! The snake might be anywhere. Already one or two revelers were declaring that it had bitten them. At last there came lights, the darkness was dispelled, and the snake turned out to be—you've guessed it—a rope! (Fawcett 1953: 99–100)

Even people without ophidiophobia are able to detect pictures of snakes embedded in neutral backgrounds faster than objects that do not create fear (Öhman et al. 2001). Öhman and his colleagues have suggested that our brains are wired for fast, preconscious visual detection of snakes. Their evidence and reasoning are consistent with the fact that the SC, LC, pulvinar, and amygdala, the basic structures of the fear module, are all involved in our ability to detect feared objects preconsciously (Dolan and Vuilleumier 2003; Liddell et al. 2005).

The SC-pulvinar visual system, involving the two visual structures of the fear module, also involves the MT and V2, where global processing of looming objects and scaled elongate objects can occur. The K pathway is also the only visual pathway known to connect the SC with the LGN and then go on to V1 where local processing and the P pathway allow scrutiny and confirmation of the presence of danger. We might think of the K pathway as being involved in a diffuse stream of "vision for detection" just before object recognition occurs via the P pathway. This would help to explain why we are so quick in low light to jump to the conclusion that we have seen a snake when it is really a rope; in dim light the K pathway should still operate but the P pathway is less functional.

Oddly, investigation into Parkinson's disease (PD) also implicates the K pathway as an important visual contributor to the fear module. Although we do not know yet what ultimately causes PD, we do know that a loss of dopamine in the brain results in disturbances of movement, balance, and fine motor control. Recall also that sufferers in the early stages of PD can correct errors in reaching and grasping by using visual feedback. When they follow their hands with their eyes they can still adjust to reach and grasp the target they want, suggesting once again that the LGN visual system can override the SC-pulvinar visual system. This is a good example of the difference between conscious and preconscious vision. Conscious vision is affected equally by the stimulus itself and by voluntary processing. By contrast, preconscious vision is stimulus-driven, automatic processing.

Despite the ability of PD sufferers to adjust using visual feedback, their early vision is adversely affected, and it is affected in ways that also imply K pathway and SC-pulvinar visual system involvement. Preconscious visual pop out occurs when salient stimuli are immediately detected despite being embedded in distractors. Sufferers of PD perform poorly compared with non-sufferers in tests designed to measure preconscious pop out. Non-afflicted people, for instance, can quickly detect lines that are oriented differently from a background of vertical lines, and they also quickly detect the letter L among + signs. Those with PD, however, detect lines oriented only at much greater angles from the vertical lines, and they need more time to detect the letter L among + signs (Lieb et al. 1999). Sufferers of PD also have slower reaction times in the presence of distracting stimuli in their peripheral visual fields (McDowell and Harris 1997), suggesting that they have a more difficult time selectively tuning out the distractors. Disrupted preconscious vision in PD is reminiscent of what happens when the SC is damaged (Bender and Butter 1987). Importantly, if this deficit in preconscious vision is a result of damage to the SC, or the pulvinar, or both, it is still likely to have its start in the K pathway because the K pathway provides the strongest visual input to both structures.

The K pathway, through its strong connections to the SC and the pulvinar and thus the rest of the fear module, also links visual deficits of PD to motor deficits of PD. One of these motor deficits resembles a typical fear response in mammals. Parkinson's disease can cause a person to freeze for no good reason. When a person freezes, the legs suddenly become immobile and stuck to the ground as the person tries to walk. Freezing often occurs at a doorway or when there are distracting obstacles in the path, as if the person cannot tune out the distractors (McDowell and Harris 1997; Nieuwboer et al. 2001). Automatic physiological responses associated with fear and anxiety occur as early as one second before freezing is detectable on videotape (Jörg et al. 2004), suggesting that the SC-pulvinar visual system and the rest of the fear module are involved. Recall that the central nucleus of the amygdala (part of the fear module) mediates the freezing response. Motor deficits have largely been associated with a loss of cells producing dopamine in the substantia nigra (Feldman et al. 1997). It has been suggested that the substantia nigra could cause freezing by disinhibiting automatic SC visual responses to distractors, particularly in peripheral vision (McDowell and Harris 1997) where the K pathway appears to be most responsive. Freezing is a natural response of mammals to objects that appear suddenly on the periphery, and the substantia nigra normally inhibits freezing and other fearful behavior in mammals via its connections with the SC. Alternatively, it has been suggested that the SC may be disinhibited by a disruption of the PPC-to-SC connection (Fielding et al. 2006). In either case, the disinhibition problem probably begins at the retina. Could it be that the loss of dopamine in the retina disrupts the input from the K pathway to the SC and pulvinar so that they are no longer very good at screening out attention to irrelevant objects, one of their main functions? Unselective information could then get transmitted to the amygdala, which is also damaged, resulting in freezing and fear around objects that normally would not be worth attending to.

Recall also that the amygdala is involved in the ability to subconsciously detect fearful faces visually. It should no longer be surprising to learn that the amygdalas of patients with PD are not responsive to fearful facial expressions (Yoshimura et al. 2005). The fear module indeed appears to be damaged by PD (see Ouchi et al. 1999).

By virtue of its strong connections to the SC and the pulvinar, the K pathway appears to be the visual pathway most strongly involved in the process of preconscious detection of predators and other dangers. That the K pathway has expanded in primates suggests that they were able to respond to selection for more reliable or faster preconscious detection of dangerous objects than other mammals. In the next chapter I will focus on how snakes, as the first and most persistent of dangerous objects, have helped to selectively modify the vision of primates living on the different landmasses of Madagascar, South America, and Africa and Asia.

CHAPTER SEVEN

Venomous Snakes and Anthropoid Primates

> But the serpent said to the woman, "You will not die. For God knows that when you eat of it your eyes will be opened, and you will be like God, knowing good and evil."
>
> *Genesis* 3:4–6

PERCY FAWCETT disappeared without a trace somewhere in South America in 1925 on what he told his wife would be his last exploration. He left behind his journals, published posthumously in 1953. In one of these journals he described his run-in with a bushmaster, a highly venomous snake.

> All of a sudden something made me jump sideways and open my legs wide, and between them shot the wicked head and huge body of a striking bushmaster. I shouted, half jumped and half fell to one side, and waited breathless for the second attack that I knew was certain to come. Yet it did not; the brute slithered down to the stream beside the trail and lay there quiet. We had no weapons with us, and as it threatened to attack again when we pelted it with stones, we left it there. It was quite nine feet long and about five inches thick, and the double fangs, if in proportion, would be over an inch in length. Experts claim that these snakes reach a length of fourteen feet, but I have never seen one so big.
>
> What amazed me more than anything was the warning of my subconscious mind, and the instant muscular response. *Surucucus* [*sic*] are reputed to be lightning strikers, and they aim hip-high. I had not seen it till it flashed between my legs, but the "inner man"—if I can call it that—not only saw it in time, but judged its striking height and distance exactly, and issued commands to the body accordingly! (Fawcett 1953: 178)

As Fawcett and many others who are drawn to the natural world would no doubt agree, snakes are frequently difficult to see, even with our convergent orbits and sharp visual acuity. Many viperids, such as rattlesnakes (*Crotalus* spp.) and puff adders, are especially cryptic partly because their coloration and typically coiled posture camouflage them nicely. In addition, many viperids are ground dwelling and are often obscured by grass and leaves. Finally, since the hunting strategy of viperids involves being still and waiting for their prey to come along, they do not readily move out of the way of approaching humans. All of these conditions work against our ability to detect them visually. Thankfully, because they would rather not be stepped on by larger creatures, some viperids have kindly evolved warning rattles and hisses to assist our excellent but not perfect vision.

The only time I have ever stepped on a snake was when I really was not looking. In 1980 I was running down a trail in Uganda's Kibale Forest (now a national park), in an attempt to keep up with a group of red colobus monkeys that was moving quickly to an area I did not know very well. My eyes were mostly focused on the monkeys in the trees, not on the ground in front of me, because I did not want to lose them. I did look ahead on the trail, too, but never down to where my feet were landing except once that I can remember. At that moment, I looked down just as a small black snake was moving across the trail right in front of me. Neither of us stopped our momentum, and as I was less acrobatic than Mr. Fawcett, my foot landed hard on the snake, and then I was past it. My reaction at the moment of contact was to expel an "uh!" from deep in my lungs. No screams, no panic. It all happened too fast. I think the snake must have been quite surprised to feel the crush of a person's boot on its back. I do not know for certain what kind of snake it was, but its coloration suggested that it could have been a forest cobra *(Naja melanoleuca)*. Whatever it was, it was nice to me; it did not even attempt to defend itself. I hope it survived. I vowed after that close call never to run in the bush again.

Twelve years later I was running in Laikipia, Kenya, forgetting that vow because Mt. Kenya was so beautiful in the colorful, fading orange hues of sunset. I wanted to get a photograph of it before the light changed. Into the glade I ran, but this time was different. Although I remembered to keep my eyes on the ground ahead of me, something about me alerted the cobra and it sensed me before I sensed it. It raised its body and flared its hooded head, ready to defend itself. Two meters away, about two running steps, my body miraculously responded and I froze before I could even articulate "cobra" in my mind. Once I regained control over my body I backed slowly away, my eyes riveted on the snake. Forgetting the photo of Mt. Kenya, I took a shaky photo of the wary snake instead, my attention captured by the rarity of seeing a fully posed cobra in front of me. That was a really close call. I have not run in the bush since.

In my fieldwork with vervets and patas monkeys on the African savannahs and savannah woodlands, I have seen snakes more often than other people who work with non-primates there. This is not because I am more alert or because I have better eyes. It is because the monkeys tell me where the snakes are by giving alarm calls and looking straight at the snake whenever they see one. Even so, there were times when no amount of peering through the dense vegetation at the base of the tree or bush revealed the snake they saw so easily. I feel safe from snakes when I am with the monkeys, but when I am in the bush without them, I am aware of my handicap and try to be vigilant. Still, I imagine I am about as good at that point as any other human and I will bet I have unknowingly walked past many snakes over the years (see also Rose et al. 2003). The tussocks of grass under the sparse canopies of trees, where humans, monkeys, and snakes all like to shelter from the sun, and the often knee-high grassy ground cover, make snakes exceedingly difficult to see.

Snakes are easier to see in the short-grass glades that dot the tree and grass-studded savannah woodlands. One cool, early morning I was walking in a glade toward the Mutara River in Laikipia, heading toward the sleeping site of one of my study groups of vervet monkeys. The glade was so open that I could have walked anywhere within it but as it happened, my trajectory took me directly on a collision course with a sunning puff adder. Again my body froze, this time a single step away from the snake and just before I thought "snake." I was beginning to notice a pattern in my behavior. Since then, I have talked to other fieldworkers who have had the same experience of freezing just before they have consciously perceived the snake. In a sense, when we freeze at the sight of snakes, we act just like rats and mice when they first detect cats, one of their main predators.

We do not have to be in the bush or forest to have an automatic freezing experience. Since I have become aware of our ability to detect potentially dangerous things before conscious awareness sets in, I have noticed the same experience with a spider that dangled in front of my face just outside my house in a suburban college town. It was not dangerous by any means, but some spiders are, and our automatic vision acknowledges it. Now I am convinced that our ability to stop in the nick of time and then proceed appropriately depending on the specific danger is not the result of blind luck but rather many millions of years of directional selection operating on the SC-pulvinar visual system (perhaps through the K pathway) for fast preconscious detection and the slightly slower P pathway of the lateral geniculate nucleus (LGN) visual system for identification. For as many tropical fieldworkers as there are spending their lives traipsing about in bush and forest, very few have been bitten by venomous snakes. One estimate places the rate of pitviper bites (all from terciopelos, or fer-de-lances [*Bothrops asper*]) in field researchers working in Central America at a low three bites per 1.5 million person-hours (Hardy 1994).

Any increase in the degree of orbital convergence improves depth perception in the lower visual field, enabling animals to break camouflage and to see more clearly what is close and in front of them. Animals with greater orbital *con*vergence are therefore expected to be able to detect and avoid snakes more reliably than animals with greater orbital *di*vergence. Because primates have among the highest degrees of orbital convergence among mammals, they should be highly adept at this skill, especially when snakes are still. My experience with the African vervets and patas monkeys confirms that they are indeed excellent detectors of snakes.

Even among primates, however, differences exist in orbital convergence, visual ability, and brain size that could affect their chances of clearly seeing what is in front of them. Anthropoid primates, of which vervets and patas monkeys are but two species, have greater orbital convergence and greater visual ability than prosimians. Anthropoid primates also have larger fear modules and associated cortical areas, and thus larger brains, than prosimians (Chalupa 1991; Barton and Aggleton 2000; Barton et al. 2003; Stepniewska 2004; Preuss 2007). Anthropoids should, therefore, be more adept at detecting snakes than prosimians.

If constrictors (or early colubroids if you cannot bear to accept a Gondwanan origin for modern mammals) were a major selective factor in the evolution of mammalian visual systems (and therefore their brains) and in the initial evolution of closer orbital convergence, visual system expansion, and expansion of associated brain structures in early primates, were they also responsible for the more extreme expression of these traits in anthropoid primates? I have to say that while it is possible, it is not probable because in order for animals to evolve larger brains they have to overcome the greater energetic cost demanded by those larger brains (Armstrong 1983; Martin 1996; Aiello et al. 2001). That is, the cost would have to be outweighed by an even greater benefit than the one that already exists. In the case of prey animals and their predators, there would have been diminishing returns of continued expansion of the brain for prey that dealt with the same predator, unless that predator changed in some way. Further visual (and brain) expansion would be favored by natural selection only if an evolutionary innovation made the predator more deadly and changed the balance of the predator-prey coevolutionary arms race.

Venomous snakes have been around since about 60 million years ago, when a subset of colubroid snakes somewhere in Africa or Asia evolved a highly potent venom delivery system (Cadle 1988; Vidal and Hedges 2002). Colubroidea is the superfamily of snakes that includes the deadly venomous snakes in the families Viperidae (vipers and pit vipers) and Elapidae (cobras, mambas, coral snakes, etc.) as well as snakes in the family Colubridae (racers, gopher snakes, kingsnakes, etc.), most of which are not deadly. Although there are some colubrid snakes, such as boomslangs *(Dispholidus typus),* that are so poisonous that they can kill people, the classically defined venomous colubroid snakes are viperids and elapids. Viperids are more ancient than colubrids and elapids (Cadle 1987, 1988; Gloyd and Conant 1990; Knight and Mindell 1994; Heise et al. 1995; Dowling et al. 1996; Keogh 1998; Lenk et al. 2001; Slowinski and Lawson 2002; Vidal and Hedges 2002; but see Gravlund 2001). Their potent envenomation innovation coincided with the appearance of fast-moving mammals such as bats, rodents, and primates in Asia or Africa, and it has been suggested that this subset of snakes adapted to the new food supply by evolving a more powerful technique to subdue their prey than mere biting and holding (Greene 1983; Cadle 1988; Feduccia 1995; Douady et al. 2002; Huchon et al. 2002; Gebo 2004).

The evidence available today suggests that both viperids and anthropoid primates evolved somewhere in Africa or Asia (Keogh 1998; Lenk et al. 2001; Beard 2002; Dagosto 2002). The earliest putative anthropoid fossils found so far have been discovered in Afro-Arabia and Asia in rocks dated to the middle Eocene (Ross 2000; Beard 2002; Kay et al. 2004; but see Ciochon and Gunnell 2002), and because it is unlikely that fossils represent the earliest of any lineage, anthropoids likely originated by at least the late Paleocene (Ross 2000; Beard 2002; Dagosto 2002; Eizirik et al. 2004; Ross and Kay 2004).

I suggest that the changes that resulted in anthropoids occurred shortly after primates arrived in Asia and snakes in the Old World evolved their highly potent venom delivery apparatus.

Prosimian versus Anthropoid Vision

Madagascar is unusual for an Old World landmass because it has constrictors and colubrids but no venomous snakes. By the time venomous snakes evolved, Madagascar was well separated from all other landmasses, and since then there has been no colonization of Madagascar by venomous snakes as has happened in Australia. Although some colubrids are venomous, none that could pose a deadly threat inhabit Madagascar

(Glaw and Vences 1994; Kardong 2002; Vidal 2002). Molecular evidence currently indicates that the prosimians in Madagascar diverged from prosimians in Africa and Asia around 70–60 million years ago (Yoder et al. 1996; Eizirik et al. 2004; Yoder and Yang 2004). If primates evolved in Indo-Madagascar, then Malagasy prosimians have never interacted with venomous snakes. If primates instead evolved in North America, Africa, or Asia, then the ancestors of Malagasy prosimians would still have had far less exposure to venomous snakes than any other primates.

I suggest that in the absence of venomous snakes in Madagascar, prosimians there have experienced no greater selective pressure beyond that created by constrictors to adapt and expand their visual systems and their brains. This would explain why even the habitually diurnal, group-living, Malagasy prosimians have visual systems that have evolved little beyond those of the earliest primates, whereas anthropoids have visual systems that have become substantially modified. Yes, the African and Asian prosimians also have simple visual systems even though they do have to deal with venomous snakes, but they are all nocturnal and this places a constraint on them that diurnal Malagasy prosimians and anthropoids do not have. As is the case for other nocturnal mammals, African and Asian prosimians would not find further expansion of the LGN visual system useful without the benefit of sunlight and so they evolved other non-visual means for avoiding venomous snakes (more on this later).

Prosimians have more divergent orbits, poorer central vision, and poorer visual acuity than anthropoids (Kay and Kirk 2000). We know they have poorer vision because, among other differences, only some prosimian species have a fovea, and when it is present it is only poorly developed (tarsiers are exceptional among prosimians in having a well-developed fovea, one reason why they are often classified with anthropoids as haplorhines) (Rohen and Castenholz 1967; Stone and Johnston 1981; Ross 2000; Kirk and Kay 2004). Less obvious but of central importance are differences between prosimians and anthropoids in metabolic activity in early vision where the superior colliculus (SC)–pulvinar and LGN visual systems coordinate with or complement each other. These differences can be seen by staining for cytochrome oxidase (CO). Staining is a common technique used to identify certain types of cells, and staining for CO identifies those cells that are highly metabolically active in the brain (Wong-Riley and Carroll 1984; Wong-Riley 1994). The darker the CO staining, the more metabolically active the neurons (Horton 1984; Allman and Zucker 1990). These highly active neurons create metabolic energy via the continuous excitatory activity of glutamate, a neurotransmitter at the synapses of neurons (Nie and Wong-Riley 1995, 1996).

Primates and some non-primate mammals with forward-facing eyes (e.g., cats and ferrets [*Mustela putoris*]) have what are called CO blobs (where CO activity is very pronounced) in V1. However, only primates have CO stripes in V2 (Livingstone and Hubel 1982; DeYoe and Van Essen 1985; Condo and Casagrande 1990; Krubitzer and Kaas 1990; Cresho et al. 1992; Lachica et al. 1993; Preuss et al. 1993; Murphy et al. 1995; Horton and Hocking 1996; Preuss and Kaas 1996; Preuss 2007) (Figure 7.1). Greater metabolic activity, as indicated by the blobs in the V1 and the stripes in the V2, may contribute to the higher energetic costs of primate brains relative to other mammals.

Although only primates have CO stripes, there is nonetheless variation within primates in the extent, clarity, or uniformity of CO staining in V2. When they are present at all, the CO stripes stain only weakly in nocturnal prosimians (no diurnal prosimians have been examined yet) (Condo

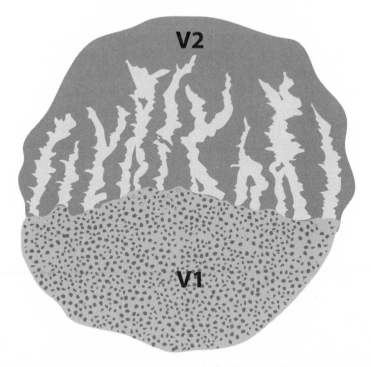

Figure 7.1   V1 blobs and V2 stripes. The darker blobs and stripes are stained for cytochrome oxidase (CO) and reveal regions of high metabolic activity.

and Casagrande 1990; Krubitzer and Kaas 1990; Preuss et al. 1993; Collins et al. 2001). In anthropoids, however, CO stripes are obvious. Even the usually nocturnal owl monkey (Fernandez-Duque 2003) has obvious stripes (Krubitzer and Kaas 1990), so the difference between prosimians and anthropoids is real and not simply the result of limited sampling of only nocturnal prosimians. The weak CO staining at best in the V2 of prosimians suggests that they may have less metabolically active brains than anthropoids.

Importantly, the blobs receive direct input only from LGN K layers (Livingstone and Hubel 1982; Hendry and Yoshioka 1994), and the pulvinar provides the major thalamic input into the darkly staining stripes (at least in anthropoids) (Livingstone and Hubel 1982; Levitt et al. 1995). Thus, the K pathway, which I have suggested is involved in preconscious "vision for detection," is directly involved with the most metabolically active parts of V1 and V2, the V1 blobs (via the LGN) and the V2 stripes (via the SC-pulvinar visual system). The peculiar involvement of the K pathway suggests that some kind of coordination between the SC-pulvinar visual system and the LGN visual system is occurring in the V2 stripes: weaker input to the V2 stripes from the LGN K layers by way of the V1 blobs is complemented by stronger input to those same stripes from the SC-pulvinar visual system (Sincich and Horton 2005) (Figure 7.2). Perhaps the K pathway provides the hypothesized circuit that mediates both automatic and attentive object-ground separation via "border ownership" neurons that have recently been found in the V2 (Qiu et al. 2007).

No one really knows why V1 blobs have greater metabolic activity. Cells in the blobs of diurnal primates are highly responsive to color, a known P-pathway attribute (Livingstone and Hubel 1982), but they must also serve another purpose because cats, ferrets, and nocturnal primates have blobs but do not have color vision as we know it (Allman and Zucker 1990). Perhaps these cells have something to do with improving depth perception. Similarly, it is not clear why V2 stripes have increased metabolic activity. It has been suggested that in addition to assisting in directing attention and screening out irrelevance in the environment, the pulvinar influences or enhances neuronal activity in V2 (Levitt et al. 1995; Soares et al. 2001a). But for what?

If metabolic activity, at least in V2, differs between prosimians and anthropoids, perhaps it is because prosimians faced less intense selection from snakes to expand their visual systems. When snake-phobic people are shown images of snakes, relative regional cerebral blood flow, a sign of neuronal activity, increases in V2 but not in V1 (Wik et al. 1993;

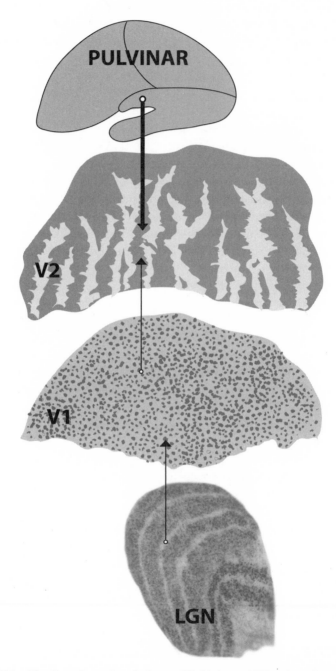

Figure 7.2   The K pathway in relation to V1 blobs and V2 stripes. Input from the K layers in the LGN target the blobs in the V1, which then send projections to the thin stripes in the V2 where they converge with input from the pulvinar.

Fredrikson et al. 1995), to which the medial and dorsal lateral pulvinar do not project. Could the greater metabolic activity in the V2 stripes help anthropoid primates perceive the totality of snakes more quickly?

While one might think that diurnal Malagasy prosimians would also benefit from having highly metabolically active visual areas to detect constrictors, remember that the increased metabolic cost of having a more active visual system might be energetic overkill against predators that have not changed over time. Moreover, energetic constraints may be even more severe for Malagasy prosimians than for other primates because their island environment is climatically harsh (Wright 1999). For instance, Malagasy prosimians are the only primates that go into torpor, or quasi-hibernation. Perhaps expanding the visual system would become more affordable for some diurnal prosimians if venomous snakes were to suddenly arrive in Madagascar and increase the cost of keeping a slightly sluggish visual system. On the other hand, perhaps they may be so energetically constrained that they would not be able to adjust and the arrival of venomous snakes would spell their doom.

Platyrrhine versus Catarrhine Vision

Although anthropoids started out living with venomous snakes, not all anthropoids have had equal exposure. When the ancestors of today's platyrrhines were still just non-specialized anthropoids in the Old World, they lived with venomous snakes. Their coexistence with venomous snakes lasted for at least 25 million years, likely enough time to modify the LGN visual system further. However, when the ancestors of modern platyrrhines arrived in South America no earlier than about 35 million years ago (Arnason et al. 1998; Nei and Glazko 2002; Schrago and Russo 2003), they arrived on a landmass that had no venomous snakes. In order for venomous snakes to arrive in South America, they first had to disperse from Asia to North America and then go on to South America (Parkinson 1999). The oldest fossil viper in South America has been dated to the late Miocene (Albino and Montalvo 2006). Molecular evidence suggests that bushmasters and lanceheads (*Bothrops* spp.) arrived sometime between 23 and 10 million years ago (Zamudio and Greene 1997; Wüster et al. 2002), but that rattlesnakes and perhaps other viperids arrived only after the Panamanian land bridge formed about three million years ago (Vanzolini and Heyer 1985; Cadle 1987; Crother et al. 1992; Greene 1997; Wüster et al. 2005). Thus, for the anthropoid primates that eventually became platyrrhines, exposure to venomous snakes was interrupted. For at least 12 million years and up to 32 million years, New World primates lived unconstrained by the selective pressure from venomous snakes. During this window of freedom, the various genera of platyrrhines diverged. The earliest radiation is thought to have begun by about 26 million years ago, and molecular evidence suggests that all the modern platyrrhine genera diverged 20–11 million years ago (Schneider et al. 1993, 2001; Chaves et al. 1999; Cropp and Boinski 2000). Multiple divergences during a time of relaxed selection would have enabled natural selection to experiment more without negative consequences, at least in relation to venomous snakes. The predicted result would be more variable visual systems among New World primates.

More variation indeed appears to exist in the visual systems of platyrrhines than in those of catarrhines. First, in the LGN, platyrrhines exhibit greater variation in the degree to which the separate P layers are intermingled (see Figure 4.5) (Le Gros Clark 1941; Kaas et al. 1978; Hendry and Reid 2000). Greater intermingling suggests greater volume and this is significant because the volume of the P layers is correlated with the degree of orbital convergence (Barton 2004).

Second, the width and intensity of CO staining in the V2 varies more within and among platyrrhine species than in catarrhines. In marmosets the CO stripes are similar in width, whereas in squirrel monkeys and owl monkeys the CO stripes form alternating thick and thin stripes (Livingstone and Hubel 1982; Wong-Riley and Carroll 1984; Tootell et al. 1983; Krubitzer and Kaas 1990). Marmosets also have lighter staining CO stripes than squirrel and owl monkeys (Kaas and Huerta 1988; Krubitzer and Kaas 1990). In contrast, although some CO stripes are conventionally called "thin" stripes because they have the same function as thin CO stripes in squirrel monkeys (the first species for which stripes of different widths were noted [Livingstone and Hubel 1982]), in both macaques and humans they are difficult to distinguish from "thick" CO stripes by width alone. The two kinds of stripes also often run together without an intervening pale stripe in both macaques and humans (Hubel and Livingstone 1987; Ts'o and Gilbert 1988; Wong-Riley et al. 1993; Roe and Ts'o 1995; Ts'o et al. 2001; Roe 2004; Sincich and Horton 2005). There also appears to be greater variation in wiring from the V2. Projections from thin stripes go to the middle temporal area (MT) in squirrel and owl monkeys but to the V4 in capuchins (Krubitzer and Kaas 1990; Nascimento-Silva et al. 2003), whereas the catarrhines that have been studied all resemble capuchins (DeYoe and Van Essen 1985; Krubitzer and Kaas 1990; Shipp and Zeki 1995). Because V2 is more binocularly sensitive than V1 (Peterhans 1997; Roe and Ts'o 1997), these differences may translate into subtle species differences in close-range depth perception and object recognition among the platyrrhines. Perhaps this is related to head cocking, a behavior suggested to function in visual exploration of novel objects for greater object discrimination (Cantalupo et al. 2002; Kaplan and Rogers 2006). Head cocking occurs rarely in catarrhines and more variably in platyrrhines. Head cocking is common in the New World marmosets, tamarins, and squirrel monkeys, for instance, but not in spider monkeys (Menzel 1980; Menzel and Menzel 1980; Cantalupo et al. 2002; Kaplan and Rogers 2006). In common marmosets *(Callithrix jacchus)*, more than 70% of head cocks occurred when the fixated object was in peripersonal space (Kaplan and Rogers 2006). Given that object recognition is a main function of the LGN visual system, there may be a relationship between head-cocking behavior and having a fairly simple LGN visual system.

Third, there is greater variation in platyrrhines within the V1. For example, in spider monkeys and capuchins, ocular dominance columns (so called because input from the eyes alternates in columns down through the several cell layers of V1) are obvious with CO staining after one of the eyes has been damaged but the other continues to function. In squirrel monkeys, however, ocular dominance columns were considered to be non-existent until very recently when better techniques to detect them became available (Hess and Edwards 1987; Florence and Kaas 1992; Horton and Hocking 1996). Individual squirrel monkeys also have variation in the clarity of their ocular dominance columns and about a third of all the individuals tested lack them entirely (Horton and Hocking 1996; Adams and Horton 2006). This variation has not been reported in catarrhines. In all catarrhines examined, including humans and the squirrel monkey–sized talapoin *(Miopithecus)*, ocular dominance columns in V1 are clearly seen with CO staining. The locations of V1 blobs are also more variable in platyrrhines than in catarrhines. Like all catarrhines that have been examined, capuchins have blobs that are consistently located relative to ocular dominance columns. Squirrel monkeys, on the other hand, have blobs that are scattered about without regard to the location of ocular dominance columns (Horton and Hedley-Whyte 1984; Hess and Edwards 1987; Florence and Kaas 1992; Horton and Hocking 1996).

Finally, platyrrhines have greater variation in color vision than catarrhines. This will be discussed more fully later on, but for now I will just mention that howler monkeys are the only platyrrhines known to be completely trichromatic, i.e., they can easily distinguish between red and green. In all other platyrrhine species that have been studied, all males and some females are dichromatic, i.e., they have what is commonly called red-green colorblindness. Some females, however, can distinguish red from green. The variability in color vision can occur even within a single group. In contrast, all catarrhines that have been studied are trichromatic, regardless of sex (Jacobs et al. 1996; Jacobs and Deegan 1999, 2001).

To date, no one has been able to explain the variability of platyrrhine visual systems. I was puzzled, too, at first, until I learned that venomous snakes did not get to South America until after platyrrhines arrived there and began their radiation. Then it made perfect sense to me.

Catarrhine visual systems may be less variable than platyrrhine visual systems because catarrhines have always coexisted with venomous snakes. They are the only primates whose genera radiated under the constraints created by venomous snakes. Greater uniformity among catarrhine visual systems is one reason why we can rely so heavily on macaques as models for understanding our own visual systems. Catarrhine primates, including humans, have the most highly developed visual systems in the mammalian world. Rather than persecute venomous snakes, perhaps we should thank them for our excellent vision.

A Genetic Correlate to Visual Expansion?

Of course, if we are talking about evolution we are recognizing that there had to have been changes in the genes of primates as a result of their coexistence with constricting and venomous snakes. I have suggested that we can explain differences between primates and other mammals in visual (and therefore brain) expansion by differential responses to snakes. Primates have greater visual ability and larger brains than other mammals, and among primates, anthropoids have greater visual ability and larger brains than prosimians, relative to body size (Barton 1999). Brain size is correlated with basal metabolic rate, particularly at birth (Armstrong 1983; Martin 1996), and the neocortex demands a lot of metabolic energy (Aiello et al. 2001).

A neat discovery is that differences in brain size mirror evolutionary changes in genes for CO. Rates of change in CO genes have been more rapid in primates than in other mammals, more rapid in anthropoids than in prosimians, and more rapid in catarrhines than in platyrrhines. In fact, expansion of the neocortex has already been suggested to have driven these changes in CO genes (Wu et al. 2000; Grossman et al. 2001; Goldberg et al. 2003). Because half of the primate neocortex is devoted to vision (Barton 1998), why not consider a link between the rate of evolutionary change in CO genes and visual systems in particular?

Cytochrome oxidase is an enzyme complex in the mitochondria that deals with electron transport, and it is critical for aerobic energy metabolism (Capaldi 1990; Kadenbach et al. 2000). Taking just V2 as an example of the possible connection between visual systems and evolutionary changes in CO genes, recall that no mammals other than primates have CO staining stripes in V2; that among prosimians, if stripes are present at all, they are not obvious; and that anthropoids have CO stripes that are more obvious. Recall also that among anthropoids, whereas some platyrrhines have CO stripes of variable width or that are only lightly stained, catarrhines consistently have darkly staining CO stripes that are mostly wide and often blend into each other. There may be other patterns of CO activity in other parts of the brain, but it is intriguing that the pattern of CO activity in V2, a visual area that all mammals have, mirrors relative rates of evolutionary change in CO genes. Even more intriguing is the fact that this cortical area of early vision where increased CO activity is clearly evident in anthropoids (1) is involved with stereopsis, figure-ground separation, occlusion, and global features; (2) has cells that respond to cues provided by snakes; and (3) has CO stripes that receive direct connections from the K pathway via both the SC-pulvinar visual system and the LGN visual system (Ogren and Hendrickson 1977; Wong-Riley 1977; Curcio and Harting 1978; Livingstone and Hubel 1982; Wong-Riley and Carroll 1984; Levitt et al. 1995; Adams et al. 2000; Sinich and Horton 2002; Stepniewska 2004).

It is often thought that directional selection is responsible for increased rates of changes in genes. Directional selection continually pushes organisms toward extremes in adaptation. When rates of gene evolution slow down after a period of fast evolution, it is often thought that directional selection gives way to stabilizing selection, in which intermediate forms are favored over extremes (Goodman 1982; Grossman et al. 2001; Wildman et al. 2002; but see Li and Grauer 1997). If this interpretation is correct, then what the rates of change in CO genes tell us is that primates have had greater directional selection than other mammals, that anthropoids have had greater directional selection than prosimians, and that catarrhines have had the most persistent directional selection of all the primates. These higher-level taxonomic differences in evolutionary change in the genes for CO in primates are consistent with the pattern of differential visual evolution in primates versus other mammals, and with a similar pattern of differential visual evolution and evolutionary coexistence of primates with venomous snakes on the different landmasses of Madagascar, South America, and Africa and Asia.

Have Venomous Snakes Ever Been a Risk to Anthropoid Primates?

These sorts of changes in the visual systems of primates likely would have occurred long ago in the last common ancestor of anthropoids. If venomous snakes were responsible for major changes in anthropoid vision, they would have had to have been major predators of anthropoids. We tend to think of anthropoid primates as being rather large, and most of today's anthropoids are certainly too large to be prey for venomous snakes (at least as adults), but when the earliest anthropoid primates evolved, they were tiny. For example, *Eosimias*, the earliest anthropoid found thus far, weighed as little as 100 grams (or less than a quarter pound), easily within the size range of the modern-day prey of venomous snakes (Greene 1997; Kingdon 1997; Beard 2002, 2004; Egi et al. 2004; Gebo 2004).

The key bit of information that started me on the path of the Snake Detection theory was a short paper that described an endogenous retrovirus in a Russell's viper from Asia (Andersen et al. 1979). We have all heard of retroviruses by now because the human immunodeficiency virus (HIV) is one that is regularly in the news. The evidence now supports the idea that humans originally acquired HIV by eating primates that carried it (Weiss and Wrangham 1999; Wolfe et al. 2004). *Endogenous* retroviruses are retroviruses that have been incorporated into the germ line and thus can be transmitted like other genetic material to progeny (Gifford and Tristem 2003). They are often evolutionarily very old and no longer a cause of mortality (Johnson and Coffin 1999; van der Kuyl et al. 2000; Gifford and Tristem 2003). The paper reported that the Russell's viper endogenous retrovirus is more closely related evolutionarily to two type D retroviruses (PO-l-Lu, or langur endogenous virus, and Mason-Pfizer monkey virus) that are carried by Asian catarrhines than to a type D retrovirus carried by platyrrhines (squirrel monkey retrovirus). Therefore, the snake likely acquired the retrovirus sometime after 35 million years ago when platyrrhines and catarrhines diverged. Indeed, the endogenous nature of the retrovirus suggests that it happened a very long time ago.

At the time the Andersen et al. (1979) paper was published, it was not understood how retroviral transmission occurs. By the time I read the paper many years later, it had been figured out that retroviral transmission typically requires some kind of physical contact. So, I had no difficulty seeing that the retrovirus could have spread from a catarrhine primate to a venomous snake through physical contact in the form of a bite from the snake. The overlapping geographic ranges of Russell's vipers with macaques and langurs, the primates carrying the related retroviruses, make it more reasonable that this was indeed what had happened. By the way, the same phenomenon likely occurred when a retrovirus that originated in primates became established at least six million years ago in felids (Wayne et al. 1989).

A common response of primates upon encountering a snake is to approach, stare, and vocalize at the snake, a response called mobbing. When vervets on the ground see snakes, for example, they stand bipedally, stare at the snake, and produce a vocalization given only in the context of snakes. This behavior then attracts others to the area until many individuals from the group are gathered near the snake (Seyfarth et al. 1980). Spectral tarsiers *(Tarsius spectrum)* sometimes also pounce on snakes when they mob (Gursky 2005, 2006), a response I have never seen from vervets. White-faced capuchins *(Cebus capucinus)* can also be more physical than vervets in their responses to snakes. They often retreat by leaving the ground for trees or moving higher into trees, but when they mob, they sometimes also drop branches onto the snake or make physical contact (Rose et al. 2003; Digweed et al. 2005). One white-faced capuchin was even observed to beat a terciopelo, or fer-de-lance, with a stick (Boinski 1988).

Snakes are so predictable in creating fear responses in monkeys that scientists often use them as the stimulus in studies of fear and the amygdala (e.g., Kalin et al. 2001; Amaral 2002). Levels of cortisol, a hormone produced during stressful events, increase when monkeys are shown snakes. This does not happen when monkeys are shown more innocuous animals, such as fish (Wiener and Levine 1992; Levine et al. 1993). It is clear that many species of primates are deeply afraid of snakes. The only logical conclusion we can derive from this is that snakes harmed primates so often in the past that the fear of snakes is now a very real part of them. Whether this fear of snakes is innate in primates, as it seems to be in ground squirrels and wood rats, or learned by observing others, is a dichotomy not really useful to consider. It is likely that anthropoids, at least, are innately primed to react to snakes with fear when they see others behaving fearfully (Cook and Mineka 1989; Öhman 2007).

Predation on primates by constrictors has been witnessed directly by fieldworkers (Chapman 1986; Cheney and Wrangham 1987; Heymann 1987; Burney 2002; Gursky 2002; Tello et al. 2002). It has also been documented by examining the stomach contents of constrictors (Shine et al. 1998). However, most primates are now too large to be eaten by venomous snakes. Has the past threat from venomous snakes continued to today?

The most voluminous evidence of envenomation of primates comes from an unpublished 13-year study by Iqbal Malik of rhesus macaques in India. She examined the bodies of nine monkeys found in the morning, dead in their sleeping trees, and determined that snakebite was the cause of death. She suspected that 35 other monkeys also died from snakebite, although she did not conduct similar postmortem investigations on them (I. Malik, pers. comm. to D. Hart; Hart and Sussman 2005). Most of the other cases come from Africa. Two female chacma baboons *(Papio ursinus)* were bitten and killed by a cobra *(Naja nivea)* in a cave (Barrett et al. 2004). An infant patas monkey that was playing with another member of its group was bitten by an undetected puff adder (Chism et al. 1984; J. Chism, pers. comm., 3 March, 2003). Patas monkeys at my study site in Kenya do not encounter pythons but do encounter and mob puff adders. The death of the infant patas monkey explains why puff adders are mobbed even though we do not consider them to be a predator of monkeys. A juvenile Syke's monkey *(Cercopithecus mitis albogularis)* was observed with bite marks from a snake, likely from a black mamba *(Dendroaspis polylepis)*. One had been seen in that same area during the past month and the juvenile expressed symptoms during the process of dying that suggested black mamba envenomation. A juvenile blue monkey *(C. mitis stuhlmanni)* died after it was bitten by a Gaboon viper *(Bitis gabonica),* which then tried but failed to eat such a large animal. As with the infant patas monkey, these juveniles were probably bitten as they played in dense vegetation (Foerster 2008). In the case of the juvenile blue monkey, the venomous snake was clearly acting as a predator. Similarly, in an attempt to escape from a pursuing black mamba, a galago fled over the shoulder of primatologist Dorothy Cheney but was still killed (D. L. Cheney, pers. comm., 4 March, 2003). Finally, in South America a buffy tufted-ear marmoset *(Callithrix aurita)* approached a jararaca *(Bothrops jararaca)* too closely and was bitten and killed (Corréa and Coutinho 1997). Collectively, these examples show that wherever primates coexist with venomous snakes, the risk of death exists, giving them good reason to continue to fear venomous snakes.

For those who would diminish my interpretation of risk simply because of the paucity of observations, I would counter with the statement that few observations of interactions of primates with venomous snakes cannot be taken to mean that the risk of death by venomous snakebite is low. After all, as Silk and Stanford (1999) pointed out, primatologists rarely observe primates giving birth but we know it happens. Determining any cause of death of wild primates is exceedingly difficult, and it is remarkable that we have any observations at all given the speed with which venomous snakes strike. Field-workers truly must be in the right place at the right time to witness such events.

For our own species, one estimate is that more than 150,000 human deaths occur per year from snakebites, and most of these are in the tropics (White 2000). People seem to be at greatest risk just before dusk and in the first few hours after dark. At those times our vision becomes poorer but we are still walking about outside (Jacobs 1993; Spawls et al. 2002). Imagine how many more people would be bitten by snakes if our eyesight was always so poor.

Do monkeys and apes distinguish between constrictors, which are clearly predators, and venomous snakes, which are mainly just deadly? Bonnet macaques *(Macaca radiata)* in southern India do appear to have this ability. They responded to models of pythons *(Python molurus)* with alarm calls, and to models of cobras *(Naja naja)* with startle and flight (Ramakrishnan et al. 2005). In contrast, different populations of white-faced capuchins respond variably to snakes but their responses are not based on whether the snakes are constrictors or envenomators. In one population, small boa constrictors and non-venomous snakes did not elicit reactions. In two other populations, however, the monkeys reacted to all snakes, and even mobbed non-venomous indigo snakes *(Drymarchon corais)* (Rose et al. 2003). The extent to which primates from the different landmasses distinguish constrictors from venomous snakes, and dangerous snakes from innocuous snakes, would be interesting to examine further.

Primatologists are quick to grant carnivorans, such as leopards, and raptors, such as crowned eagles, very strong powers to affect the lives of primates. Suggested evolutionary responses of primates to carnivorans and raptors have included large body and canine size, group living, solitary living, and multi-male coexistence in groups (Alexander 1974; van Schaik 1983; Terborgh and Janson 1986; Cheney and Wrangham 1987; Dunbar 1988; Isbell 1994b, 2004; Isbell and Van Vuren 1996). By comparison, snakes are given a nod only rarely, and generally only in the context of the evolution of alarm calls. Why should we not also acknowledge a powerful effect on primates from snakes? They have been around far longer than the other types of predators. Granted, they may not need to eat as often as the others, but they have higher population densities (e.g., Macartney et al. 1988). And even though venomous snakes do not now eat the larger primates, evolutionarily it does not matter. Whether a primate is eaten or just bitten, it is still just as dead.

CHAPTER EIGHT

Why Only Primates?

So when the woman saw that the tree was good for food, and that
it was a delight to the eyes, and the tree was to be desired to make
one wise, she took of its fruit and ate; and she also gave some to
her husband, and he ate. Then the eyes of both were opened.

Genesis 3:6–7

THE ANCESTOR from which all therian mammals descended is assumed to have been active at night, a time period that limits the
full expression of what eyes can do. Mammals, primates included, thus started out with olfactory systems stronger than their visual
systems. Given all the advantages of excellent vision, we have to wonder
why it was that only primates evolved heightened visual abilities, including sharper visual acuity and better depth perception, which allow them
to easily see objects that are close against the background near to them
and to find camouflaged objects. If snakes were predators of all mammals early on, and vision helps animals detect their predators, why did
other mammals not evolve along the path taken by primates? It is not because primates were the only diurnal mammals; the changes in their visual systems began while primates were still nocturnal. Fossils of early
nocturnal primates called omomyoids show that their visual systems
were more specialized and primate-like than those of other mammals,
and today's nocturnal primates have visual systems that are similar to
those of diurnal primates and different from those of other mammals.

I think I found the answer when I learned that with the exception of
bats there is typically a trade-off between vision and olfaction in mammals (Barton et al. 1995; Gilad et al. 2004). (For bats, the trade-off is between vision and echolocation instead, and I will discuss this further
later on in the book.) Visual systems simply cannot expand much in
mammals that cannot afford to weaken their reliance on olfaction (or
echolocation). What might have determined which mammals would have
to maintain a strong sense of smell and a weaker sense of sight and which

could afford to take the new path of reduced olfaction and expanded vision once snakes began eating them? One critical factor may have been the kinds of foods mammals ate.

From a generalized early mammal, today's 18 orders of modern placental mammals eventually arose as animals settled into different environmental and dietary niches. Although the earliest mammals would not be considered fearsome in the way that *Tyrannosaurus rex* or today's carnivorans are, they were predators, nonetheless, and probably presented a formidable danger to insects and other small animals. Have you ever taken the time to smell an insect? Try smelling a butterfly or a ladybug. To our relatively insensitive noses, most insects and other arthropods do not have much of an odor. Even the cockroaches, earwigs, and weevils that we associate with unclean and smelly places do not have a scent that we can easily detect. Some do produce noxious odors, but only when they are threatened. Despite the weak scents of insects and other invertebrates, mammals with stronger olfactory systems than ours easily sniff them out. Earthworms, beetles, ants, termites, and so many other prey items of small insectivorous mammals live hidden from sight by soil and leaf litter, and yet their predators can cut through what to us smells only overpoweringly earthy or musty and find their foods.

When angiosperms, the flowering plants, first appeared about 140 million years ago (probably in Gondwana) (Morley 2000), they opened up new dietary niches for animals. Like the conifers and other gymnosperms from which they evolved, angiosperms produced leaves and seeds. Some animals exploited this by becoming leaf and seed eaters, i.e., predators of plants. The very narrow, fibrous, and unpleasant-tasting leaves (e.g., pine needles) and the seeds encased in hard, inedible cones of conifers are examples of gymnosperm protective adaptations to minimize damage from such predators. Angiosperms also protect their leaves and seeds from predation, loading them with chemicals that are toxic to animals. This is why there is nicotine in tobacco and coca leaves, opium in the unripe seedpods of poppies, and cyanide in apple seeds and cherry pits. Most leaves and seeds, like insects, also have very little odor or are noxious to us. Imagine how hard it would be for an individual from an olfactorily oriented species to find its food if the food did not produce a strong odor and that individual happened to be born, like us, with a weaker sense of smell. Because eating better than others is a prerequisite for better reproductive success, such animals would naturally be at a selective disadvantage. Animals that were insect or plant predators simply could not afford weaker olfactory systems, even if they could benefit from visual expansion. Even today, there still exist predators of both animals and plants with small, laterally placed eyes (Ravosa and Savakova 2004).

Mammals that could not afford to switch from olfactory dominance to visual dominance would not have been easy pickings for snakes, however, because their strong olfactory sense also helps them avoid predators. Rodents, for example, can detect the scents of snakes and other predators, and avoid areas visited by those predators (Dell'Omo and Alleva 1994; Randall et al. 1995). After venomous snakes evolved, some mammal species also evolved physiological resistance to snake venoms. Most of the mammals known to be immune to snake venoms are common prey for snakes (e.g., shrews, mice, rats, voles, and squirrels), but others are carnivorans (e.g., mongooses) that are known to be snake predators (see Daltry et al. 1996; reviewed in Pérez and Sánchez 1999). This resistance would have been stronger than what was already in place for dealing with toxins from other animals such as stinging insects (Metz et al. 2006).

Like vision, venom resistance is undoubtedly beneficial in the right environment but is expensive to acquire and maintain. Thus, in areas where rattlesnakes and ground squirrels coexist, ground squirrels have venom resistance, but in areas where rattlesnakes are non-existent, ground squirrels have no venom resistance, and when previously exposed ground squirrels move to rattlesnake-free areas, their populations begin to lose venom resistance. Molecular studies indicate that it takes about 9,000 years for the descendants of ground squirrels previously exposed to rattlesnakes to lose a little over half of their resistance, and by 60,000 years, venom resistance is almost gone (Poran et al. 1987; Coss et al. 1993).

Even mammals that are not typically thought of as being resistant to snake venoms may survive better the longer their ancestors spent coexisting with venomous snakes. For example, domestic cats in Australia are more likely than domestic dogs to survive untreated snakebites (Mirtschin et al. 1998). Fossil and molecular evidence indicate that the ancestors of domestic cats originated in Eurasia (Martin 1989) and thus have coexisted with venomous snakes since their divergence from other carnivorans. The ancestors of dogs, however, diverged from other carnivorans in North America during the Oligocene (Martin 1989), before venomous snakes arrived on that continent from Asia (Gloyd and Conant 1990). According to one study, the body part of dogs bitten most often by snakes is the head (Hackett et al. 2002). This is not surprising because the head is the leading body part in dogs and they often walk along, sniffing as they go, with their heads low to the ground where they are most likely to encounter snakes. On the other hand, if their eyes were as good as ours, they would probably see the snakes before their heads were within striking distance.

Whereas our superior vision allows us to detect even motionless snakes, non-primates (and perhaps prosimians) may depend heavily on movement in order to reliably see snakes. California ground squirrels have no problem seeing moving rattlesnakes but they often do not see immobile ones within a cluttered environment of small stones, short grasses, and other vegetation (Coss and Owings 1985). Lavinia Grant nicely describes the inability of domestic dogs to see immobile snakes.

> The other day they streamed past a huge puff-adder which was lying like a log on a patch of bare ground. Its head and neck were doubled back along its body, ready to strike. Wasp, Sambu's daughter, passed within two feet of it. I called her away, but sensing from my voice that I had seen something exciting and dangerous, Wasp turned back to investigate. To my horror she passed within striking distance of the snake again without seeing it. It never moved! (2001: 70)

Wasp and Sambu are typical of relatively visually deficient non-primate mammals. Indeed, this visual weakness is exactly what sit-and-wait predators like rattlesnakes and puff adders have evolved to exploit.

Angiosperm Reproduction and Vision

When angiosperms evolved they did something that was different from the gymnosperms. Instead of using abiotic forces such as fire, water, or the wind to help them reproduce, they began to use animals. These angiosperms evolved flowers, nectar, and fleshy fruits enclosing still-protected seeds to entice some animals to pollinate other flowers and to disperse their seeds. The plants gained because animals are more accurate and efficient than abiotic forces such as wind in placing pollen among flowers, and more predictable than abiotic forces such as fire in casting seeds away from the parent tree. As a result, natural selection favored the evolution of increasingly attractive reproductive parts. To primates, this means that fruits and flowers have become sweeter and more strongly and pleasantly scented over evolutionary time. Being good primates ourselves, we humans are so drawn to sweet fruits that we have deliberately bred our domestic fruits to be far sweeter than the wild stock from which they came.

Visual expansion became possible with both the evolution of pleasantly scented, tasty fruits and flowers and the exploitation of these foods by some vertebrates, including birds and mammals. Unlike animal and plant predators, however, those non-flying mammals that took advantage of this new food source could afford a weaker olfactory system without jeopardizing their ability to find food. Since there was no foraging cost associated with olfactory reduction, their visual systems were no longer constrained. Today, the mammals with the best vision tend to be those with diets heavily weighted toward fruits. These include primates, the most frugivorous order of mammals in existence today, and bats, coming in at a distant second.

The biblical story recounted in the epigraph of this chapter about Eve's frugivory in relation to vision may have been the first to recognize a link between diet and visual expansion but it does not explain the mechanism. What has frugivory got to do with good vision? One hypothesis is that it takes more brain power to remember the locations of fruits than to remember the locations of leaves because fruits are more unpredictable or variable in time and space (Clutton-Brock and Harvey 1980; Milton 1988). Another hypothesis that more directly addresses vision is that frugivory favored better vision via expansion of the P pathway. The P pathway, if you will recall, is involved with color vision and the P layers of the lateral geniculate nucleus (LGN) are most complex among the uniformly trichromatic catarrhines. The argument is that primates with trichromatic color vision, i.e., those with the ability to distinguish reds and oranges from greens, were favored by natural selection because trichromacy enables primates to more easily find ripe fruits against a background of green foliage, or more easily detect fruit quality and ripeness, than dichromacy (which does not distinguish between reds and greens) (Polyak 1957; Mollon 1989; Barton 1998, 1999; Sumner and Mollon 2000a; Regan et al. 2001; Smith et al. 2003). Trichromatic color vision would be useful only to diurnal primates, of course, because colors cannot be seen without light.

Frugivorous primates do indeed have more complex P pathways and larger brains than leaf-eating primates. Even among closely related genera this holds true. For example, frugivorous spider monkeys have larger brains than the more folivorous howler monkeys, and frugivorous chimpanzees have larger brains than the more folivorous gorillas (*Gorilla* spp.) (Barton 1998, 1999, 2000). But these hypotheses assume that the correlation is causal, i.e., that frugivory was the main selective pressure directly favoring visual and brain expansion. Frugivorous bats suggest that this assumption may be wrong.

Two taxa of bats have evolved away from an insectivorous diet to a fruit-based diet, but interestingly, only Old World fruit bats have expanded their vision significantly. In addition to being the mammals whose superior colliculi (SCs) most closely resemble primates, they have large eyes, and most of them have lost their ability to echolocate. Indeed, before molecular studies ruled it out, many people thought bats and primates were closely related because of the similarities in their visual systems (Figure 8.1). Recall that these fruit bats were used as evidence against the Visual Predation hypothesis and as evidence supporting the hypothesis that frugivory (along with insects) was important in visual expansion (Martin 1986; Sussman 1991). There is a problem, however, that the Angiosperm/Omnivore hypothesis did not address. The New World leaf-nosed bats (the frugivorous phyllostomids), the other frugivorous bat taxon, have eyes that are much smaller than those of Old World fruit bats (Baron et al. 1996). While it is true that they have the largest eyes of all the microchiropterans, they have retained the echolocational ability of their ancestors to locate objects in their environment (Jones and Teeling 2006) (Figure 8.2). These neotropical leaf-nosed bats reveal that frugivory may be necessary but not sufficient to explain visual expansion.

The Role of Frugivory in Primate Vision

Why should mammals switch from eating insects, which have high protein levels, to eating fruits and flowers? What do they get out of such foods? How could eating fruits and flowers make it possible for vision (and the brain) to expand? Here I present an alternative that views the role of frugivory as permissive rather than selective.

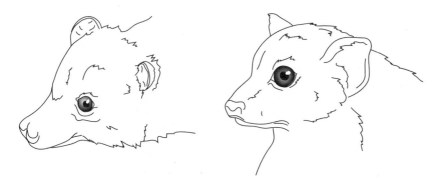

Figure 8.1 Comparison of the face of a Mariana fruit bat (an Old World fruit bat) on the left with that of a ring-tailed lemur on the right. The visual systems of most Old World fruit bats are convergent with those of primates.

Figure 8.2   Faces of a New World leaf-nosed bat on the left and an Old World fruit bat on the right. Note the size of the eyes relative to the ears.

The greatest difference in brain size between primates and other mammals occurs before birth. For any given body weight, primates have twice as much brain tissue as most other mammals (Martin 1990). Brains can be highly metabolically active tissues, and primates have at least two visual areas where neurons are uniquely highly metabolically active. Recall that high CO activity reflects high levels of neuronal metabolic activity involving glutamatergic excitatory activity. Glutamate is the main excitatory neurotransmitter in the central nervous system, and it is necessary (Orrego and Villanueva 1993). However, having too much glutamate can be bad. When neurons are exposed to high amounts of glutamate, the cellular environment can become toxic and cells can degenerate as a result of something called excitotoxicity (Lucas and Newhouse 1957; Olney1969, 1990; Choi 1988; Meldrum and Garthwaite 1990). Although our bodies naturally produce glutamate, we can increase the amount we have in our bodies by ingesting it. Glutamate excitotoxicity was first discovered, in fact, by giving young mice monosodium glutamate (MSG). This is the same flavor enhancer that we put in our foods and that gives many people headaches. Too much MSG resulted in damaged retinas in the mice (Lucas and Newhouse 1957; Olney 1969).

How does glutamate excitotoxicity occur? One way is when cells do not get the supplies they need to make energy, which mainly comes from glucose (Hertz et al. 1999). If there is insufficient glucose or CO the cells cannot make enough energy, the effect being that more glutamate will be released presynaptically to act repeatedly on its postsynaptic receptors

(Novelli et al. 1988; Henneberry 1989; del Rayo Sánchez-Carbente and Massieu 1999; Brown and Bal-Price 2003). Persistent opening of the ion channels causes more calcium ions to pour into the cells, and too many calcium ions causes a further malfunction of energy metabolism and the release of even more glutamate (Choi 1988). Then the cells die (see also LeDoux 2002).

Because glutamate is an amino acid product of the metabolism of glucose (Feldman et al. 1997), we also get glutamate indirectly from foods that have glucose in them. Ripe fruits and flowers are examples of glucose-rich foods (Romano et al. 1993; Riba-Hernandez et al. 2005). Before you stop eating your recommended daily servings of fruit for fear of glutamate excitotoxicity, however, rest assured that glucose-rich foods are not a problem unless something else is wrong, such as diabetes. In fact, glucose normally has just the opposite effect—it is a neuroprotectant. In other words, glucose helps protect neurons from glutamate overexposure (Henneberry 1989; Romano et al. 1993; Seo et al. 1999; Guyot et al. 2000).

The neuroprotectant property of glucose might help explain the permissive influence of frugivory on visual and brain expansion. The hypothesis is that as females began to eat fruits, and plants began to make fruits more attractive, a diet richer in glucose could have initiated a positive feedback loop in which greater consumption and more rapid metabolism of sugars by mothers both allowed greater CO activity to occur during development of fetal visual systems and other parts of the brain because more glucose provides more energy (and more efficiently than protein and lipids, which need to be converted to glucose), making it possible for metabolic and glutamatergic activity to increase, and was required because glucose is a neuroprotectant against increased glutamate exposure. Over evolutionary time, this could have resulted in greater neural growth, resulting in more complex visual systems and larger brains. Today, glutamate has widespread effects throughout the brain. It is heavily involved in many kinds of learning, for instance (Haberny et al. 2002). Two effects that are relevant here are that glutamate enhances fear-related learning in the amygdala (Walker and Davis 2002) and enhances learning in color discrimination tasks (Popke et al. 2001). Glutamate is also crucial for neuronal growth during development (Haberny et al. 2002).

If glucose is important for fueling and protecting more expansive visual systems and larger brains, then we should expect frugivorous primates to prefer fruits that are higher in sugars. In fact, spider monkeys (*Ateles geoffroyi*) do select fruits with higher sugar levels while passing up fruits on the same tree with lower sugar levels (Riba-Hernandez et al. 2003). We know from our own experience that ripe fruits are sweeter than unripe fruits. We also know that ripeness can often be determined by color (Sumner and Mollon 2000b). Fruits that are red or orange when ripe are typically more glucose-rich than other individuals of the same plant or species that are still green and unripe (Sanchez et al. 2000; Iglesias et al. 2001). Additionally, some plant species produce redder fruits than others, and it turns out that in the Central American forest where the spider monkeys were studied, redder fruits have higher concentrations of glucose (Riba-Hernandez et al. 2005).

The ability to distinguish reds and oranges from greens helps primates choose ripe, sugary, glucose-rich fruit over less ripe, less glucose-rich fruits. Not all primates have that ability, however. Whereas all catarrhine primates can distinguish reds from greens and are trichromatic, all platyrrhines, except howler monkeys (more on them soon), are either trichromatic or dichromatic. Thus, all males and some female platyrrhines are dichromatic and cannot distinguish red hues from green but some females can (De Valois and Jacobs 1968; Jacobs 1995). Two subspecies of prosimians also have variable trichromacy (red ruffed lemurs [*Varecia variegata rubra*] and Coquerel's sifakas [*Propithecus verreauxi coquereli*]), but in all other prosimians tested both sexes are dichromatic (Tan and Li 1999; Jacobs et al. 2002). Trichromacy has been widely thought to confer an advantage over dichromacy in finding ripe fruits. This is supported by one study in which trichromatic individuals of two tamarin species (saddleback tamarins [*Saguinus fuscicollis*] and red-bellied tamarins [*S. labiatus*]) learned to find and eat ripe, red fruits faster than dichromatic individuals in captivity (Smith et al. 2003).

In my own fieldwork, as part of a manipulative study of the foraging behavior of vervet monkeys, I placed within the vervets' visual range foods that they had never encountered before. For a long time, I could not get them to "see" raisins as food and I was about to give up when my field assistant suggested using carrots instead. When I placed a slice of carrot in with the pile of raisins, they immediately approached and inspected the food. Once the carrot was used as "bait" the vervets quickly learned to eat the high-sugar raisins, but they never ate the carrots. In retrospect, I surmise that the monkeys might not have been attracted to the raisins initially because they looked like impala droppings. In any case, this example is useful because it shows that frugivorous animals go for sugar even when it is dissociated from color. The color of fruits may be important for finding foods but it is the glucose in those foods that really matters.

Neural expansion is expensive; it does not happen simply because it can. The neotropical fruit bats show that frugivory is not enough. There had to be a selective pressure that affected catarrhines differently than platyrrhines, and anthropoids differently from Malagasy prosimians, because their visual systems are different. Recall that catarrhines have the most complex LGN P layers, platyrrhines tend to have less complex P layers, and prosimians have even simpler P layers. The P pathway is involved in central vision, particularly for distinguishing objects near and in front of oneself. It is also involved in trichromatic color vision. I have argued that the K pathway, via both the SC-pulvinar visual system and the LGN visual system, helps primates detect snakes better, perhaps by improving pop out from the background. The K pathway has direct neural connections to the CO blobs in the V1 and the CO stripes in the V2, and I have argued that increased CO activity in the blobs and stripes further improves pop out. Importantly, because primates vary in the number of LGN P layers, and because LGN K layers separate P layers, primates also vary in the number of layers devoted to the K pathway in direct relation to the number of P layers present. This suggests that the P and K layers co-evolved and expanded together.

Indeed, expansion of the K pathway, along with increased CO activity in the V1 and V2, likely would not have been able to occur unless the neurons involved were somehow buffered against glutamate excitotoxicity. One buffer is calbindin, a protein that is associated with the K pathway. Calbindin binds the calcium ions before they can cause a problem (Montje et al. 2001; Rodman et al. 2001). Another buffer, as I previously mentioned, is glucose. The P pathway facilitates increased intake of glucose via its role in color vision.

To summarize, I am suggesting that frugivory made it possible for visual systems (and brains) to expand in primates because the glucose in fruits protected the K and P pathways from glutamate excitotoxicity as the pathways expanded. The K pathway expanded under selection to better detect snakes preconsciously while the P pathway expanded in concert, to evaluate the K pathway's initial response, to help fuel visual expansion, and to protect it during expansion. The addition of trichromatic color vision to the P pathway was a later contribution that enabled frugivorous primates to find the more glucose-rich foods more efficiently.

Recently, an argument was made that trichromatic color vision evolved not to distinguish fruits from foliage but to distinguish young red leaves from mature green leaves (Dominy and Lucas 2001). The importance of having high levels of glucose readily available to protect against glutamate excitotoxicity suggests that a diet of young leaves could not have permitted visual expansion and larger brains because leaves are not as high in glucose as fruits. Moreover, other visual cues such as size and shape are available to help distinguish between young and mature leaves. Nevertheless, perhaps a leafy diet could have favored trichromatic color vision in some cases (Lucas et al. 1998) when it was not necessary to have a large brain at the same time. This may have been the case with howler monkeys. They are the only platyrrhines of which all individuals are trichromatic (Jacobs et al. 1996; Kainz et al. 1998; Regan et al. 1998). Howler monkeys are more folivorous than frugivorous (Crockett and Eisenberg 1987), and more folivorous than many platyrrhines with variable trichromacy. Perhaps trichromacy evolved in howlers to distinguish live, healthy leaves, which are typically green, from dead and dying leaves, which range from yellows and oranges to reds and browns. Trichromacy evolved independently in howler monkeys and catarrhines (Hunt et al. 1998; Kainz et al. 1998; Surridge et al. 2003), and so the selective pressure favoring the evolution of trichromacy may not be the same for both.

Certainly, trichromacy did not result in howlers having larger brains. Howlers have very small brains for their body size (Harvey et al. 1987) and they have a reputation among primatologists as being among of the most lethargic animals in the primate world. Both characteristics suggest low metabolic activity. In fact, their dearth of activity has already been attributed to their largely folivorous diet (Milton 1980; Terborgh 1983; Crockett and Eisenberg 1987). Trichromacy in the relatively small-brained howlers suggests that, as with frugivory, trichromacy per se did not drive brain expansion.

The Anthropoid Shift

The K and P pathways really expanded with the anthropoids. Most anthropoids today are active only in the daytime when good vision is particularly useful. How did primates invade this diurnal niche? I have not found any hypotheses that attempt to explain this shift, but I can think of one scenario and it is consistent with the Snake Detection theory.

When venomous snakes appeared in the Old World, they provided the selective pressure for vision to become the predominant sense for the primates that also lived there. Directional selection from venomous snakes continued to favor orbital convergence to improve stereopsis in near space so that the first tiny, nocturnal, largely frugivorous proto-anthropoids could see venomous snakes better. With visual expansion came a reduction in olfaction, which led to changes in circadian rhythms. We know olfaction is connected to circadian rhythms because when the olfactory bulbs of rodents and mouse lemurs are removed, weakening their sense of smell, the animals immediately become more active during daylight hours. Rodents and mouse lemurs also undergo a change in circadian rhythm of body temperature (Perret et al. 2003). Thus, the change is an endogenous change, a change that happens within their bodies, and not a result of the difference in lighting. In other words, rodents and mouse lemurs were not more active because they could see their insect food more easily in daylight.

It is not difficult to imagine the same kind of shift occurring slowly over evolutionary time as a much more gradual decline in olfactory ability occurred. It could have caused a positive feedback loop to develop in which the trade-off between visual gain and olfactory loss led to increased daytime activity, which drove further visual gain and olfactory loss, and so on. The end result of this positive feedback loop would have been the anthropoids: largely frugivorous, highly visual, olfactorily diminished, diurnal primates.

Among the anthropoids, the selective pressure from venomous snakes was most persistent for those that remained in the Old World. Of all the primates, catarrhines have been under the most unrelenting pressure to be able to detect venomous snakes ahead of any step, reach, or grasp. Of all the primates, catarrhines have had the greatest visual expansion and greatest olfactory reduction. In addition to having the most voluminous and complex P pathway, the most expanded K pathway, uniform trichromacy, and the largest brains of all the primates, catarrhines have lost a number of olfactory receptor genes that platyrrhines (except howlers) still retain (Yokoyama and Yokoyama 1989; Gilad et al. 2004). The reality of a trade-off between vision and olfaction is further reinforced by howlers, who have also lost some olfactory receptor genes (Gilad et al. 2004).

In a twist to the story, adding trichromacy to the P pathway to fuel the expanding visual systems and protect catarrhines from glutamate excitotoxicity would have also introduced a side effect: a weakened ability to break through camouflage (Morgan et al. 1992; Buchanan-Smith et al. 2005; Saito et al. 2005). This would have been a problem because many venomous snakes are well camouflaged. The solution would have been even greater orbital convergence, and indeed, catarrhines have the closest orbital convergence of all primates (Barton 2004). (While it is true that many insects are also well camouflaged, because catarrhines are not more insectivorous than other primates, their greater orbital convergence would not have been a consequence of selection to find camouflaged insects.)

Finally, this scenario implies that olfactorily guided locating, reaching for, and grasping food (as occurs in rodents) would have been replaced by visually guided locating, reaching for, grasping, and manipulating food as olfaction weakened. When did this shift occur? We know that diurnal frugivorous prosimians have smaller olfactory systems than nocturnal frugivorous prosimians but that all prosimians still have fairly well-developed olfactory systems (Barton et al. 1995). They also still use olfaction (and hearing) to some degree to get their food (Charles-Dominique 1977; Sussman 1991; Crompton 1995). In fact, it has been argued that the olfactory systems of prosimians are so similar to those of other mammals and so unlike those of anthropoids that they should not even be considered primates (Cave 1973). As I mentioned earlier in the book, a common assumption is that all primates employ visually guided reaching and grasping for obtaining and manipulating their food, but that assumption needs to be tested. While some prosimians clearly do use vision to obtain their food, others do not appear to rely so heavily on it. The scenario I have just laid out suggests that the shift to visually guided reaching for, grasping, and manipulating food certainly would have been in place in the last common ancestor of the anthropoids, if not before.

CHAPTER NINE

Testing the Snake Detection Theory

This is the most absurd, stupid, ridiculous "scientific" conjecture, dream, fantasy, BS crap I have ever seen. Obviously, some people will believe ANYTHING, and obviously, some "scientists" are on powerful drugs that send them off into never-never land on wild, speculative, and idiotic paths of their imagination. So much absolute crap passes for "science"!

Labeling someone's after-too-much-pizza nightmare as "science" is just another way to try to propagandize the whole evolutionary figment of imagination. This is absurdity at its highest level.

What hogwash. This is a prime example of why academics should be kept on a short leash.

This is nothing more than a guess with absolutely nothing to support it.

Yahoo! News online discussion

W HEN THE scientific paper from which this book was developed was described in the news media, I was amazed at the volume of attention it received and the reactions of many people. No other theory on the origin of primates or the origin of human vision has inspired reports by news services all around the world. It was covered by the press not only in the United States, where the work was done, but also in countries as culturally diverse as China, Malaysia, India, and Australia. It also showed up on numerous Internet personal blogs. I was even asked to write an op-ed piece about it for the *New York Times* (Isbell 2006c).

As much as I would like to believe that we are all suddenly interested in evolution, I think it is more likely that the theory generated widespread public interest only because snakes were involved. Snakes bring out strong reactions in a lot of people. In the *Yahoo! News* online discussion group, for example, many people shared the sentiment that it was the most ridiculous theory they had ever heard and a waste of taxpayers' money. Other people took a different tack and complained that the theory was so obvious that taxpayers' money should not have been wasted on it. I do not know where the complaints about taxpayer money stem from because I never actually used taxpayers' money to formulate this theory. Typical statements are offered at the beginning of this chapter to give you a sense of the emotional vehemence expressed by so many anonymous people toward the Snake Detection theory.

Perhaps the Snake Detection theory hit a nerve because it connects humans and evolution, two words that do not go together in the same sentence for some people. Perhaps other people had an adverse gut reaction because the Snake Detection theory sounds a lot like the story of Eve and the Serpent, and maybe they thought it was developed to support the creation story common to the Abrahamic religions (Islam, Judaism, and Christianity). I, myself, enjoy imagining that the person who originally came up with the story of Eve and the Serpent was simply a very good naturalist. Regardless of our emotional response to the Snake Detection theory, if we hope to test it further, we will have to proceed logically and systematically.

How can the Snake Detection theory be tested? It will be challenging, to be sure, because it is impossible to know for certain what happened millions of years ago. As Darwin showed so well, the best that can be done with evolutionary theories is to examine them from every available angle. If the evidence builds against the theory over time, then we can feel more confident that life really did not happen that way and we can look for other explanations. If the evidence cannot tear it down, then we can feel progressively better about the veracity of the theory.

In this chapter, I will begin the process of testing the Snake Detection theory. I hope you will see that there are enough lines of evidence, indirect though they may be, for us to proceed with efforts to examine the Snake Detection theory further. My greatest hope is that it will inspire you or someone else to pick up where I have left off. This is an interdisciplinary theory that will take time to test fully, one that could benefit from input from a host of scientists: herpetologists, ornithologists, primatologists, other mammalogists, molecular geneticists, immunologists, neuroscientists, behavioral ecologists, psychologists, geologists, paleontologists, and evolutionary ecologists. Whatever your interest, there is room here for many people with many talents.

Old World Fruit Bats, Trees, and Caves

Old World fruit bats have been used as evidence against the Visual Predation hypothesis and as evidence in favor of the Angiosperm/Omnivore hypothesis (Sussman 1991). As I mentioned previously, Old World fruit bats are the mammals that most closely resemble primates in the retinal projections to the superior colliculus (SC) (Pettigrew et al. 1989; Rosa and Schmid 1994), and I suggested that this convergence is the result of similar environmental conditions: roosting in lighted areas in the daytime where both bats and early primates would be more exposed to predators but better placed to take advantage of vision to detect and avoid them.

Rousette fruit bats *(Rousettus aegyptiacus)* are, like other Old World fruit bats, frugivorous and nocturnally active, and they use both vision and olfaction to locate their foods (Baron et al. 1996; Kalko et al. 1996). They are exceptional among Old World fruit bats, however, in that they are the only ones that are still able to echolocate (Heffner et al. 1999). This suggests that they have not made as great a trade-off between vision and echolocation as other Old World fruit bats. Examination of the visual system of rousette fruit bats also suggests that it is exceptional among the Old World fruit bats that have been studied. Instead of having a more primate-like SC than other mammals, as other Old World fruit bats have, rousette fruit bats have a non-primate-like, typical mammalian SC (Thiele et al. 1991). The Snake Detection theory predicts, therefore, that we will find that rousette fruit bats do not roost in sunlit places during the day. In fact, this is what is found. Rousette fruit bats do not roost in trees and other sunlit places but rather roost in caves and other dark places (Baron et al. 1996). Comparative studies provide some of the strongest tests of evolutionary hypotheses, and fruit bats may be a particularly good group of non-primate mammals with which to explore visual evolution in primates further, especially in relation to the SC.

Given the similarities in vision between primates and some Old World fruit bats, it might be tempting to draw the conclusion that there is a causal relationship between bat vision and snakes, as the Snake Detection theory does for primates, because the bats with the best vision are found in the Old World where venomous snakes evolved. This would probably be wrong, however. Although snakes were undoubtedly predators of early bats, just as they are today (Greene 1997; Hammer and Arlettaz 1998), bats do not have the same paleobiogeographical history with snakes that primates have. Bats are members of Laurasiatheria and evolved about 64 million years ago in the northern continents (Teeling et al. 2005). Moreover, while venomous snakes evolved from a colubroid snake in Africa or Asia no later than 60 million years ago, Old World fruit bats did not evolve until about 24 million years ago (Teeling et al. 2005), and by that time *all* the major predators of bats were already present (Rydell and Speakman 1995). Old World fruit bats roosting in trees would have had to keep vigilant not only for snakes but also for raptors and carnivorans.

Why Do Snake-Eating Raptors Have Such Big Eyes?

In an update of his Visual Predation hypothesis, Cartmill (1992) wrote that "adaptive explanations must be general enough to predict similar adaptations in other cases and they must be rejected if those predictions are not borne out" (107–108). Although Old World bats shed light on the evolution of the SC-pulvinar visual system, the combination of dietary preadaptations and selective pressure from snakes was so unique to primates that no other mammals can do justice as models for reconstructing the whole of primate visual evolution. On the other hand, some birds may be great models. First, they are so distantly related to primates that any similarities are likely to be independently evolved, not just passed on from a common ancestor. Second, many birds are diurnal and have excellent vision. Third, both birds and primates have large eyes relative to body weight (Brooke et al. 1999; Ross 2000; Kirk and Kay 2004). Fourth, the diets of many birds are broadly similar to those of primates. Finally, birds are eaten by snakes, and snakes are eaten by birds (Greene 1997; Weatherhead and Blouin-Demers 2004; Webb and Whiting 2005), so the potential exists for snakes to have acted as selective pressures on birds. I am not suggesting that snakes were the driving force behind all the visual adaptations of both birds and primates. Birds also did not have the same pattern of coexistence with snakes that primates had. Still, there may be parallels worth exploring further.

Consider raptors, for example. Although anthropoids have the greatest visual acuity of all the mammals, diurnal raptors surpass them (Ross 2000; Kirk and Kay 2004). Raptors also have fairly convergent orbits (Edelstam 2001), and there is no doubt that they need good visual acuity and close-range stereopsis to pounce on their prey. They have, in fact, been used as models supporting the Visual Predation hypothesis (Ross 2000).

Raptors are also excellent for testing the Snake Detection theory. They are diurnal visual predators, and some have become specialized as snake predators (Brown 1976; Ferguson-Lees and Christie 2001). The Snake Detection theory predicts that snake-eating raptors will have larger eyes (for greater visual acuity) or greater orbital convergence (for improved stereopsis) than raptors that are not snake specialists and do not have as pressing a need to detect snakes.

As predicted, although raptors generally have larger eyes for their body weight than other birds (Brooke et al. 1999), the eagles and falcons that specialize in eating snakes have even larger eyes than other raptors (Edelstam 2001; Ferguson-Lees and Christie 2001). Snake specialists also have larger heads than other raptors (Edelstam 2001; Ferguson-Lees and Christie 2001), presumably because larger heads can house larger eyes. One only has to look at a field guide to see that African snake eagles (*Circaetus* spp.) have larger eyes than other African eagles, with large, owl-like heads and necks (Brown 1976; Zimmerman et al. 1999; Ferguson-Lees and Christie 2001) (Figure 9.1). Another snake-eating specialist, the South American laughing falcon *(Herpetotheres cachinnans)* has "obviously large eyes and head (owl-sized but not especially owl-like)" (Ferguson-Lees and Christie 2001: 812) (Figure 9.2). The larger eyes of snake eagles and laughing falcons indicate greater visual acuity because larger eyes mean greater visual acuity among diurnal birds and mammals (Kiltie 2000; Ross 2000; Kirk and Kay 2004). Similarly, raptors that are snake-eating specialists are considered by some to have greater binocular vision than other diurnal raptors, a trait expected to be particularly valuable because of the great precision that is required when handling such dangerous prey (Edelstam 2001). On the other hand, others do not see much of a difference (Martin and Katzir 1999). This will need to be examined further.

Figure 9.1   African snake eagle (left), a snake-eating specialist, compared with an African fish eagle (right), a non-snake-eating specialist. Note the snake eagle's larger eyes and head.

Figure 9.2   The laughing falcon, a snake-eating specialist from South America. Note the very large eyes and head. Convergence of these traits in snake-eating raptors on separate continents suggests that snakes may have favored changes in their visual systems, as also hypothesized for primates.

Two distantly related taxa of snake-eating raptors on two different continents that both have larger eyes for increased visual acuity (and possibly greater binocularity) than other raptors that are not snake-eating specialists clearly satisfies Cartmill's demand. They provide two independent, non-primate cases of visual adaptation to snakes. By the way, because the Nocturnal Visual Predation hypothesis predicts that convergent orbits are not required of diurnal visual hunters (Cartmill 1992), diurnal raptors such as hawks and eagles do not actually work as models for that hypothesis.

Do Primates Differ in Their Reactions to Snakes?

We know that Malagasy prosimians have never been exposed to venomous snakes, that platyrrhines have had exposure on and off and on again, and that catarrhines have had the longest continuous exposure. This should translate into different reactions to snakes among the primates, with catarrhines showing the strongest reactions and prosimians the weakest. Platyrrhines might be intermediate and perhaps more variable in response than catarrhines or prosimians. A recent study suggests, in fact, that some Malagasy prosimians do not get nearly as excited as catarrhines do in the presence of snakes (even model snakes). Old World monkeys and apes uniformly react fearfully to both model and live snakes, complete with agitation, alarm calling, and intense visual monitoring of the snakes (visual monitoring apparently being so reliable a cue from mammals in general that spitting cobras have evolved the defensive mechanism of aiming their venom at the eyes), but when wild brown mouse lemurs *(Microcebus rufus)* were exposed to model snakes, some did not act perturbed in the least. Some even walked on the models as if they were just another branch (Deppe 2005, 2006).

The first of such comparisons was actually conducted more than 100 years ago. British scientists P. Chalmers Mitchell and R. I. Pocock (1907) described their observations of variation in the responses of primates from Madagascar, the New World, and Africa and Asia to live snakes.

We wish first to record the extremely interesting fact that Lemurs differ markedly from true Primates, inasmuch as they exhibit no fear of snakes whatever. It was most curious to notice how, when we approached adjoining cages, the one with lemurs the other with monkeys, carrying with us writhing snakes, how the monkeys at once fled back shrieking, whilst the lemurs crowded to the front of the cage, displaying the greatest interest and not the smallest perturbation when a snake was brought so close to them that its tongue almost touched their faces. We got the impression that had the lemurs been given the opportunity, they would at once have seized and tried to devour the snake. The South American monkeys showed fear in irregular and sometimes slightly marked form. Spider-monkeys *(Ateles)* were quite as excited and alarmed as any Old World monkey. Some of the larger Cebidae did not retreat, but uncovered their canines and looked as if they were ready to show fight. Some small specimens retreated but showed no special alarm, others were nearly indifferent. The Old World monkeys of all the genera in the Society's Collection recognised the snakes instantly and bolted panic-stricken, chattering loudly and retreating to their boxes or as high up as possible in the larger cages. Our large Baboons, including the huge Mandrill, were even more panic-stricken, jumping back in the greatest excitement, climbing as far out of reach as possible and barking. Of the Anthropoids, the Gibbons were least timid; one small agile Gibbon *(Hapale agilis)* [sic] showed no fear and very little curiosity; a larger one of the same species and a Hoolock receded but without showing panic. It is possible that the very markedly arboreal habits of the Gibbons have brought them so much less in contact with snakes that their fear of snakes is partly obliterated. The Chimpanzees, except one baby which was indifferent, recognized the snakes at once and fled backwards, uttering a low note sounding like "huh, huh." They soon got more excited and began to scream, getting high up on the branches or on the wire-work of their cages, but all keeping their eyes fixed on the snakes. (793–794)

Further studies should be conducted to examine the prediction more systematically.

Do Primates Differ in Their Ability to Detect Snakes?

We know that differences exist in the visual systems of primates. Most obviously, prosimians have less orbital convergence and poorer visual acuity than the anthropoids. Do visual differences translate into measurable behavioral differences in the ability to detect snakes? One prediction is that anthropoids can detect snakes more reliably or more quickly from farther away than can prosimians. This prediction is currently being tested by one of my graduate students.

Do platyrrhines and catarrhines differ in their ability to detect snakes? How much do platyrrhines depend on movement to detect snakes? In captivity, snake-naïve cotton-top tamarins *(Saguinus oedipus)* do appear to use movement as a cue for potential danger. In one experiment, the tamarins showed fearful reactions equally to a moving rat and a snake, and more to a moving snake than to a still snake (Hayes and Snowdon 1990). On the other hand, South American titi monkeys *(Callicebus moloch)* seem well able to detect immobile snakes. Several years ago I conducted a short study to determine the cues used by captive titi monkeys to detect snakes. They were clearly able to see an immobile model snake under their cages. Intriguingly, regardless of the color of the snake model or how well it matched the background of model vegetation, they responded most strongly when the snake models were dressed in netting that resembled the lines created by snake scales.

Interestingly, for their body size, patas monkeys have the largest eyes of any primate measured, and Kirk (2006a) attributed this distinction to their high cursoriality. At first glance, this seems reasonable given their long legs and their reputation as among the fastest of primates. However, a systematic study of locomotor activity budgets revealed that patas monkeys spend about as much of their time running as sympatric vervets, a closely related species that does not have unusually long legs. Instead, how they differ is in their walking: patas monkeys spend much more time walking while foraging than vervets (Isbell et al. 1998). I suggest that we consider an alternative: patas monkeys' more terrestrial habits are likely to expose them more frequently to puff adders or other terrestrial and well-camouflaged venomous snakes. Like the large eyes of snake-eating raptors, the large eyes of patas monkeys could be particularly useful for seeing under such conditions.

How Do Nocturnal African and Asian Prosimians Deal with Venomous Snakes?

Unlike Malagasy prosimians, African and Asian prosimians live with venomous snakes. Because of the constraints of limited light, vision in the nocturnal African and Asian prosimians could not have evolved to the extent that it has in diurnal anthropoids in response to snakes. The Snake Detection theory, therefore, predicts that, like many non-primate mammals, Asian and African prosimians have instead evolved non-visual defenses against venomous snakes.

In fact, a number of odd traits do occur only in Asian and African prosimians that could be used as defenses against snakes. For instance, when slow lorises bite, they inject an allergenic substance that causes pain and even anaphylactic shock and death in humans (Krane et al. 2003). Pottos have large scapulas (shoulder blades) and prominent, spiky vertebrae just below their necks. When they are threatened, they curl down their heads and aim the vertebrae toward their attackers (Charles-Dominique 1977). Such a defensive position might deter snakes because it makes the prey much harder to swallow head first, the common position of consumption of vertebrates by snakes (Greene 1997). Of course, this would not matter if the potto was envenomated and died without being eaten. Thus, another strategy of pottos is to plummet to the ground from the forest canopy when they are confronted by snakes (even model snakes) (Charles-Dominique 1977), an anti-snake strategy not reported for any other primate except the angwantibo *(Arctocebus calabarensis),* another nocturnal African prosimian (Charles-Dominique 1977).

Tarsiers are often used to support the Nocturnal Visual Predation hypothesis because they are nocturnal and insectivorous, with the largest orbits of any mammal of their body size (Cartmill 1992; Ross 2000). Attributing the tarsier's unusual visual adaptations entirely to insectivory may be premature, however, because they sometimes eat venomous snakes (Niemitz 1984). The exceptionally large eyes and convergent orbits of snake-eating raptors suggest that animals that eat venomous snakes must have large eyes for the visual acuity required to quickly distinguish the head from the rest of the body even when the snake is coiled up, and the stereopsis to precisely judge the distance to the snake's head when they attack. This might also apply to tarsiers. For such snake-eating animals, the selective pressure must have been intense. With most prey items, one miss and all that is lost is a meal. When the prey item is a venomous snake, one miss could mean losing one's life. It would be worth exploring whether tarsiers have venom resistance, as do other animals that eat venomous snakes. It would also be worth examining whether all African and Asian prosimians have greater venom resistance than anthropoid primates.

Why Are the Few Cathemeral Primates in Madagascar?

It has been hypothesized that some Malagasy prosimians are cathemeral, i.e., able to be active both day and night, because they are in evolutionary disequilibrium now that humans have caused diurnal primate-eating raptors and numerous lemur species in Madagascar to become extinct (van Schaik and Kappeler 1996). The argument is that the extinction of these predators has recently freed up some formerly nocturnal prosimian species to expand a bit further into the daytime niche. This hypothesis generated a lot of interest but has not held up as new evidence has come in. For example, we now know that there are still primate-eating raptors in Madagascar (Colquhoun 2006). In addition, Kirk (2006b) has shown that the corneas of cathemeral lemurs are intermediate in size between those of diurnal and nocturnal lemurs and that they are also comparable in size to other cathemeral non-primate mammals. The fact that all the species of the genus *Eulemur* are cathemeral and have intermediately sized corneas indicates that cathemeral lifestyles are not recent innovations and have been around for least 12–8 million years since their last common ancestor (Yoder and Yang 2004). We currently have no good explanation for the distribution of cathemerality in primates.

The Snake Detection theory might fill this gap. It predicts that venomous snakes, having acted as a selective pressure favoring primates' specialized visual systems, diminished olfaction, and diurnality, should have been especially deadly to those primates with more generalized visual systems, i.e., vision useful under both nocturnal and diurnal conditions. Cathemerality, for whatever other reasons (e.g., as a response to competition or to climatic extremes), would be possible only where venomous snakes are not a threat. This prediction is supported by the observation that cathemerality is common only in Madagascar (Fleagle 1999) where there are no venomous snakes.

If it ever becomes possible to distinguish among cathemeral, diurnal, and nocturnal species in the fossil record, another prediction would be that cathemeral species will decline in the fossil record after venomous snakes appear in Asia or Africa.

Why Are There No Habitually Terrestrial Platyrrhines?

Platyrrhines appear to be less willing than catarrhines to come close to the ground, and also seem to spend more time scanning the ground. Increasing wariness as primates come closer to the ground thus far has been documented in capuchins, woolly monkeys, titi monkeys, tamarins, and langurs *(Presbytis)* (van Schaik and van Noordwijk 1989; Müller et al. 1997; Steenbeek et al. 1999; Di Fiore 2002; Miller 2002; Prescott and Buchanan-Smith 2002). All but the langurs live in the New World. This cautious behavior has been interpreted as an indication of high risk of predation by mammalian carnivores, but if this is the case, then why are there so few Old World species on this list? I know from unfortunate experience that leopards can devastate Old World monkey populations (Isbell 1990; Isbell and Enstam 2002), and there are numerous other reports of predation by carnivorans on primates from other Old World study sites (e.g., Busse 1980; Hoppe-Dominik 1984; Stanford 1989; Boesch 1991; Tsukahara 1993; Zuberbühler and Jenny 2002; Henschel 2005; D'Amour et al. 2006). Snakes might alternatively explain the difference in wariness and reluctance to come to the ground.

Indeed, captive Geoffroy's marmosets *(Callithrix geoffroyi)* delayed going to the ground to forage for insects in the morning after they had been exposed to a snake model the previous evening. The marmosets also increased the number of times and length of time they inspected the area where the snake had been placed (Hankerson and Caine 2004). No comparable work has been carried out yet with catarrhines. The Snake Detection theory predicts that greater resistance of platyrrhines to terrestriality is correlated with poorer vision, perhaps especially pop-out vision. (The absence of venomous snakes in Madagascar would also help to explain why Malagasy prosimians often do come down to the ground.) A second prediction of the Snake Detection theory is that time spent on the ground is positively correlated with visual expansion in New World monkeys. Thus, for example, capuchins, which have among the most complex P layers of the platyrrhines, should spend more time on the ground than other platyrrhines. A third prediction is that we should find terrestrial platyrrhines in the fossil record before venomous snakes arrived. In fact, it has been suggested that 26-million-year-old *Branisella* was more terrestrial than present-day platyrrhines based on its tooth shape and wear, and the semi-arid or arid environment in which it lived (Takai et al. 2000).

Neuroscientifically Relevant Research

Investigation of neural differences might also provide insights. One might, for example, quantify baseline activity of cytochrome oxidase (CO) or cerebral blood flow in primates from the different landmasses and then compare the extent of short-term change in CO activity or cerebral blood flow in visual areas or the fear module after exposure to snakes and more neutral objects, such as tree branches and flowers. In light of the known responsiveness of neurons in inferotemporal cortex (IT) to faces and other complex and salient images, it might also be interesting to examine whether single cells in visual areas further along the ventral stream also respond vigorously to snakes.

Other Possibilities

The evidence presented not only in this chapter but also throughout the book is consistent with the Snake Detection theory, which suggests that snakes contributed to (1) the expansion of the visual systems and the fear module in mammals, (2) even greater expansion of the fear module along with greater connections to vision in primates, and (3) the pattern of variation in vision within the primates. In this book, I have offered one potential explanation for the origin of primates, the origin of anthropoids, and the origin of differences between catarrhines and platyrrhines.

Some readers may balk at such an all-encompassing theory. Others may appreciate that although snakes could have been important, they were not the only important factor, i.e., that two (or more) good reasons to evolve a trait are better than one. This conceptual camp believes we cannot disentangle selective pressures to identify any one as the primary selective pressure for the trait we are considering. This book will likely rankle people in that camp. Another conceptual camp believes, however, that for many traits there had to be a first, or at least primary, selective pressure. I am obviously in this latter camp, and I have presented evidence that snakes were the first of the major predators of primates and were thus primarily responsible for the expansion of the visual systems in primates. I am not saying that other pressures had no effect. For instance, I have already mentioned that once primates started living in groups, there would have been additional advantages to having good vision. Almost certainly, however, group living in primates appeared after the changes in vision had begun to evolve.

In support of the hypothesis that the need to avoid snakes was ultimately responsible for the unique visual systems of primates, and, to some extent, for variation in the visual systems within primates, I have relied on both proximate (e.g., neurological and behavioral) and evolutionary (e.g., molecular systematics and paleontological) levels of inquiry, all the while couching everything in functional terms (e.g., survival, reproductive success, and natural selection). For those who just want the bottom line, or who are totally confused by all the details, Figure 9.3 condenses the theory to its main steps and conclusions.

Table 9.1 suggests many ways to test the Snake Detection theory should you want to pursue this line of questioning further. In proposing the Snake Detection theory, I have specified some essential predictions that would fail to support the core of the theory if they were refuted. I have also suggested a number of more minor predictions that would not cause rejection of the core theory if one or another is refuted.

While all current competing hypotheses are broad in the sense that they attempt to explain the entire suite of primate characteristics, they limit themselves to addressing only the origin of primates. They do not address why it is that primates themselves differ in their visual systems. They also do not describe the conditions or preadaptations that might have been necessary in order for primates to have taken the evolutionary route toward visual expansion, why other mammals did not also take that route, and the steps involved in the evolution of greater and greater visual expansion. The Snake Detection theory does all this. Hypotheses that are more comprehensive in their explanation of a trait are more useful, all else being equal. The Snake Detection theory also makes sense of other patterns that have not been adequately explained yet, including why some primates are cathemeral in Madagascar, why primates in the New World are not habitually terrestrial, and how and why anthropoid primates became diurnal. The Snake Detection theory explains why there is variation among the visual systems of primates, and how this variation evolved.

"The study of evolution proceeds from observed facts and found data more than from controlled experimentation" (Quammen 2006: 176). This is how I have proceeded. I have noted facts and found data in fields ranging from molecular genetics and neuroscience to paleontology and plate tectonics to investigate an idea born out of my own specialties, behavior and ecology. I have used the terminology available to us at this

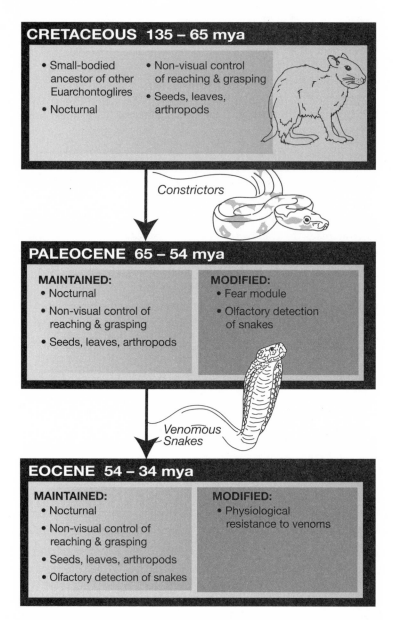

Figure 9.3a Summary of the Snake Detection theory. Although non-primate mammals were exposed to both constrictors and venomous snakes, they did not have the kind of diet that favored visual expansion. Some eventually evolved physiological resistance to snake venoms; mya = million years ago.

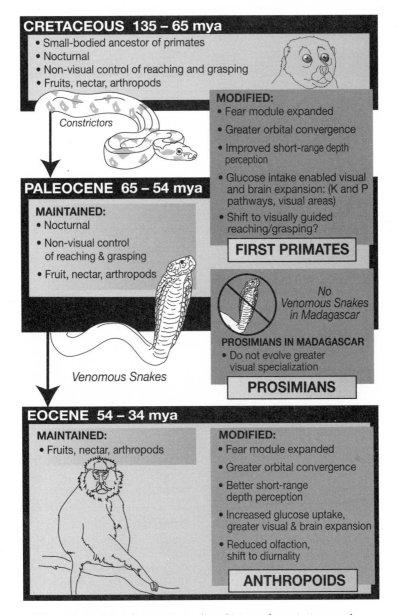

CRETACEOUS 135 – 65 mya
- Small-bodied ancestor of primates
- Nocturnal
- Non-visual control of reaching and grasping
- Fruits, nectar, arthropods

Constrictors

MODIFIED:
- Fear module expanded
- Greater orbital convergence
- Improved short-range depth perception
- Glucose intake enabled visual and brain expansion: (K and P pathways, visual areas)
- Shift to visually guided reaching/grasping?

FIRST PRIMATES

PALEOCENE 65 – 54 mya

MAINTAINED:
- Nocturnal
- Non-visual control of reaching & grasping
- Fruit, nectar, arthropods

No Venomous Snakes in Madagascar

PROSIMIANS IN MADAGASCAR
- Do not evolve greater visual specialization

PROSIMIANS

Venomous Snakes

EOCENE 54 – 34 mya

MAINTAINED:
- Fruits, nectar, arthropods

MODIFIED:
- Fear module expanded
- Greater orbital convergence
- Better short-range depth perception
- Increased glucose uptake, greater visual & brain expansion
- Reduced olfaction, shift to diurnality

ANTHROPOIDS

Figure 9.3b The origin of primates and evolution of prosimians and anthropoids from the Cretaceous to the Eocene as a result of differential evolutionary exposure to constricting and venomous snakes. A diet high in sugars enabled brain expansion to occur in primates in response to the appearance of constrictors. While prosimians in Madagascar have never been exposed to venomous snakes, and remain with small changes in the mammalian visual system to this day, primates that lived in India, Asia, and Africa faced an even greater risk from venomous snakes beginning in the Paleocene. As a result, their visual adaptations were modified further and today they are known as anthropoid primates.

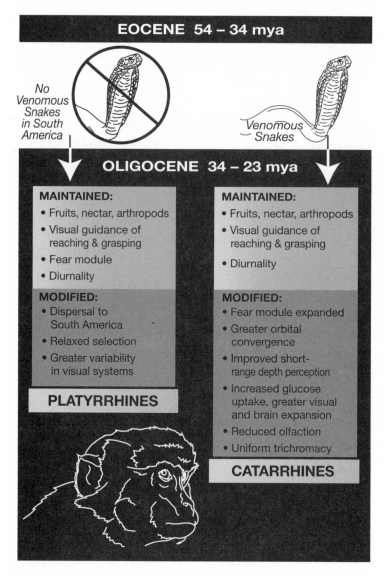

EOCENE 54 – 34 mya

No
Venomous
Snakes
in South
America

Venomous
Snakes

OLIGOCENE 34 – 23 mya

MAINTAINED:
- Fruits, nectar, arthropods
- Visual guidance of reaching & grasping
- Fear module
- Diurnality

MODIFIED:
- Dispersal to South America
- Relaxed selection
- Greater variability in visual systems

PLATYRRHINES

MAINTAINED:
- Fruits, nectar, arthropods
- Visual guidance of reaching & grasping
- Diurnality

MODIFIED:
- Fear module expanded
- Greater orbital convergence
- Improved short-range depth perception
- Increased glucose uptake, greater visual and brain expansion
- Reduced olfaction
- Uniform trichromacy

CATARRHINES

Figure 9.3c   During the Oligocene, some anthropoids dispersed to South America where they did not encounter venomous snakes (constrictors were there already). There they diversified, and in the absence of venomous snakes as a continuing selective pressure, platyrrhine visual systems become more variable. In contrast, anthropoids that remained behind in the Old World continued to be exposed to venomous snakes, and as a result, their visual systems became more impressive in acuity, depth perception, and color vision. Today the catarrhine primates have the best vision of all mammals.

Table 9.1   Predictions of the Snake Detection theory.

Core theory:

Snakes acted as the primary selective pressure operating on primates to expand their visual systems. Varying evolutionary exposure to venomous snakes resulted in differences in visual capability within primates.

	Condition		
Core predictions:	S	N	O
Catarrhines have the most specialized visual systems of the primates.	X		
Catarrhines detect snakes faster, or more reliably, or from a greater distance than platyrrhines and prosimians.		X	
Platyrrhines have more variable visual systems than catarrhines (not an artifact of sampling bias).	X		
Anthropoids detect immobile snakes faster, or more reliably, or from a greater distance than prosimians.		X	
Prosimians have simpler visual systems than anthropoids.	X		
Venomous snakes evolved in Asia or Africa before catarrhines radiated.	X		
Venomous snakes arrived in South America after platyrrhines arrived.	X		
Platyrrhines radiated in the absence of venomous snakes.	X		
Venomous snakes (viperids and elapids) have never existed on Madagascar.	X		

	Condition		
Corollary predictions:	S	N	O
A. Behavioral differences that reflect differences in visual ability:			
Visually guided reaching and grasping evolved in the last common ancestor of anthropoids, if not before.		X	
Anthropoids rely less on movement than prosimians to detect snakes.		X	
Catarrhines can detect partially occluded snakes more quickly, or more reliably, or from a greater distance than platyrrhines.		X	
Absence of habitual terrestriality in platyrrhines is a result of poorer ability to detect snakes on the ground.		X	

(continued)

Table 9.1   *(continued)*

Corollary predictions:	Condition		
	S	N	O
Non-primate mammals rely more than primates on snakes' movement to detect them.		X	
Dichromatic primates can see camouflaged snakes more reliably than trichromatic primates.		X	
Nocturnal prosimians in Africa and Asia have greater venom resistance than other primates.		X	
Cathemeral primates were largely selected against outside of Madagascar because they were poor at detecting venomous snakes (and those that are cathemeral outside Madagascar have little contact with venomous snakes).		X	
B. *Cell response properties:*			
V1 CO blobs enhance stereopsis and depth perception.		X	
V2 CO stripes enhance pop out of objects against a cluttered background.		X	
Cells in the V2 are more responsive than cells in the V1 to global attributes of snakes.		X	
Cells in the V1 are more responsive than cells in the V2 to detailed attributes of snakes.		X	
Cells in common visual systems of mammals are sensitive to unique attributes of snakes, e.g., scale patterns; elongate, moving shapes.		X	
C. *The SC-pulvinar visual system:*			
The unique retinal connections to the SC in primates and Old World fruit bats are a result of selection on their vision to better detect predators while roosting in sunlit places.		X	
The medial pulvinar of primates has expanded to assist them in avoiding predators, especially snakes.		X	
The SC-pulvinar visual system controls automatic, visually guided reaching and grasping in primates.			X

Table 9.1   (continued)

Corollary predictions:	Condition		
	S	N	O
The SC-pulvinar visual system is involved in preconscious predator detection and avoidance.	X		
The SC-pulvinar visual system has expanded in primates in response to snakes as predators.		X	
The primate K pathway is primarily involved in "vision for detection," i.e., automatic, preconscious vision.		X	
The primate K pathway is most developed in catarrhines, less so in platyrrhines, and least in diurnal prosimians.		X	
D. *The LGN visual system:*			
The LGN visual system is involved in conscious vision.	X		
The primate LGN and SC-pulvinar visual systems are integrated through the K pathway and evolved initially to detect, identify, and assess the risk from snakes.		X	
The primate P and K pathways co-evolved.		X	
The primate P pathway is most developed in catarrhines, less so in platyrrhines, and least in diurnal prosimians.	X		
The P pathway expanded in part to fuel the expansion of the K pathway (which expanded to better detect snakes).		X	
The P pathway expanded in part to protect the brain from excitotoxicity while expanding to detect venomous snakes more reliably.		X	
The neuroprotectant property of dietary glucose permitted an increase in metabolic activity in the brains of primates.		X	

(continued)

Table 9.1   (continued)

	Condition		
Corollary predictions:	S	N	O
Catarrhine trichromacy evolved to improve efficiency in finding foods with high glucose content in response to selection to detect snakes quickly and reliably.		X	
The earliest primates ate a diet high in fruits, flower, and nectar.	X		
Trichromacy in howler monkeys evolved to distinguish live from dead leaves.		X	
E. *The evolution of CO genes:*			
CO activity, CO gene evolution, and evolutionary time of coexistence with venomous snakes covary in primates.		X	
CO gene evolution is related to visual expansion in primates.		X	
CO gene evolution is related to neocortical expansion in primates.			X

Core predictions are those that must be confirmed in order to support the Snake Detection theory. Corollary predictions are those that would support the theory if they are confirmed, but would not necessarily falsify the theory if they are refuted. Under the category *Condition,* predictions with direct or indirect evidence are supportive (S). Other predictions have been developed in the context of the Snake Detection theory (N) or suggested by others in a different context but are still applicable to testing the theory (O). They are suggestions for future investigation.

point in our species' history to develop the Snake Detection theory. The theory may seem novel because of the modern scientific jargon and the amazing, new techniques that have allowed us to probe brains as never before, but as the story of Eve and the Serpent shows us, it is not such a novel idea. Whatever the terminology we have available to us at any point in time, we always seem to come back to the same observation: snakes and primates have had a very long history together, and there have been consequences for primates. Primates clearly reveal their long history through their fear of snakes. The Snake Detection theory presents evidence that primates also reveal their long history with snakes through their vision.

Implications for Humans

I N ADDITION to the overlying theme that snakes were responsible for the evolution of our exceptional vision, there is an underlying theme to this book, which is that if there has been any "progression" in the evolution of animals, it has been toward greater and greater behavioral flexibility in the animal's responses to its environment. For primates, the main sensory interface with the environment is vision. The changes in primate vision have resulted in their having greater control over their own lives. By way of gathering together the main points of the preceding chapters, let me reinforce how these two themes intertwine.

Vision, Snakes, Choices, and Decisions

The visual system that all vertebrates have is the superior colliculus (SC)–pulvinar visual system. It is a visual system that allows vertebrates to shift their attention to important and unexpected objects in the environment without the need for any assessment or conscious awareness. Rapid detection of danger, for example, is critical; assessment would only slow things down. One of the functions of the SC-pulvinar visual system is to facilitate freezing, a common response to visual detection of fast approaching objects. Many vertebrates have difficulty seeing objects that remain motionless, so freezing is generally a good strategy for minimizing predation (LeDoux 2002).

The lateral geniculate nucleus (LGN) visual system was an innovation of mammals that gave them the ability both to perceive and assign intent and to override the SC-pulvinar visual system when the information gathered allowed it. In conjunction with the SC-pulvinar visual system, which draws attention, for instance, to that scaly thing in the grass, the LGN visual system gives mammals the ability to determine whether that scaly thing is really a snake or just the head of a dead lizard. While the SC-pulvinar visual system detects the object and directs attention to it, the LGN visual system, with its widespread connections to the neocortex, enables mammals to identify the object as a snake (or not), ignore it, move away from it, utter alarm calls and stare at it, attack it, or eat it. The LGN visual system gave mammals more choices, and with those choices came the responsibility of more decisions. Mammals are able to perceive, scrutinize, and determine the level of risk that snakes (and, once snakes appeared, other creatures) present, and act accordingly.

Primates were no different from the other small mammals that had to avoid snakes except that they roosted in places exposed to the light, they had a glucose-rich diet, and they had a remarkable pattern of coexistence with snakes. As was also the case for some of the Old World fruit bats, roosting in sunlit places was conducive to visual modification of the SC-pulvinar visual system to detect and avoid predators. If fruit bats had evolved at about the same time as primates, they might have at least evolved the kind of vision that nocturnal prosimians have, but they evolved only about 24 million years ago, not 80–90 million. Beyond their antiquity, what was so unique about primates was their pattern of evolutionary coexistence with venomous snakes. Visual systems are more developed in those primates that have shared the longest evolutionary time with venomous snakes and least developed in those primates that have had no exposure at all to venomous snakes. Somewhere between the two extremes lie the visual systems of primates that have had intermittent evolutionary exposure to venomous snakes.

A diet high in fruits and nectar freed primates from dependence on olfaction to find their foods and permitted the metabolic and neurological expansion of primate visual systems. The SC-pulvinar visual system expanded largely by increasing the size of the pulvinar in the thalamus while the LGN visual system expanded by increasing the number of P layers in the LGN, also in the thalamus. Both visual systems were tied together by an expanded K pathway. The K pathway may be involved with the preconscious detection of predators and other threats, and its expansion may mean that, among other attributes, primates are better than other mammals at preconsciously detecting non-moving predators visually. Anthropoids especially are expected to be quite good at seeing objects pop out against a cluttered background. Expansion of the P layers gave anthropoids greater visual acuity, protected neurons from excitotoxicity, and fueled expansion of the cortical areas involved in vision, which allowed anthropoids to perceive in even greater detail and assess ever more subtle differences in their environments, including subtle changes in their social environments.

The ability to perceive and assess subtle visual differences means more choices that require more decisions. For example, frugivorous primates have to decide whether one fruit is riper and more worth eating than a different fruit, and they have to do this over and over again every day. In the social domain, primates have to decide whether others' attempts at social engagement are friendly or antagonistic and then they have to decide how best to respond. Thus, a primate can choose to approach another, groom her, appease her by lip-smacking, act as if she does not see her, walk away, threaten her, or avoid her, among other options (Byrne and Whiten 1988; Whiten and Byrne 1988; Dunbar 1992, 1998, 2003).

These considerations mean that life has become more complicated for primates than for a lot of other animals, but there is a positive side. They have gained flexibility, the ability to change their behavior when situations call for rapid adjustments. Primates have gained greater control over themselves. It reminds me again of the story of Eve and the Serpent. Once the snake appeared, Eve paid attention to the snake and made her own decision to eat the fruit. I like thinking that that part of the story, the moment when Eve decides to eat from the Tree of Knowledge, is a greater than 2,000-year-old attempt to explain our ability to perceive, assess, and respond to novel situations. That moment in the story has been vilified in modern times, and perhaps for good reason. While behavioral flexibility has been good for us from an evolutionary perspective (otherwise we would not be here now), we are each also responsible for the decisions we make, and we have not always made the best decisions.

Our human vision is very much like that of macaques, and I have argued in this book that our excellent vision is mainly the result of evolutionary pressure to detect and avoid snakes. Many of us also share with macaques and other primates a fear of snakes. We know the fear is irrational in most cases but we cannot stop feeling it. Perhaps phobias arise when something in our brains gets mixed up and the LGN visual system, with its heavy cortical involvement, fails to override the SC-pulvinar automatic visual system. The inability to control how we would like to behave seems very much like the absence of cortical control, or the failure of the cortex to override the SC-pulvinar visual system. As an example of cortical failure, persons with lesions in certain parts of the posterior parietal cortex (PPC) can reach for objects accurately only when they are looking directly at them. They cannot dissociate their reaching from their gaze, and cannot carry out multiple actions simultaneously (Jackson et al. 2005). Similarly, others with PPC lesions cannot look to places where they are not pointing (Carey et al. 1997). Cortical involvement allows us to disengage the automatic hand-eye connection generated by the SC-pulvinar visual system unless the situation requires urgency of response, in which case the SC-pulvinar visual system steps up.

If snakes had stopped being a problem for our primate ancestors, natural selection probably would have let us relax a little and we probably would not have ophidiophobia today. Alas, their effect on us continues to modern times. And because snakes still exist in rural parts of Africa, we can be fairly certain that snakes also affected our hominin ancestors there. Indeed, viper fossils in Africa have been dated to the Miocene (Lenk et al. 2001), revealing that snakes were a danger to our ancestors even before hominins became bipedal.

From four to six million years ago, our ancestors took a trajectory quite different from other primates when they began to walk upright. It is worth investigating how they might have dealt with snakes once they became bipedal and their eyes were raised higher from the ground. With their eyes farther from the ground it would have been more difficult to see the highly cryptic and often motionless venomous snakes that live in the savannahs where they walked.

The Natural History of Declarative Pointing

We often acknowledge language and complex tool construction as two behaviors that set humans apart from other animals. There is another behavior, however, that also appears to distinguish us from other animals. Ours is the only species that points declaratively (Povinelli and Davis 1994; Butterworth et al. 2002; Tomasello et al. 2007; but see Leavens et al. 2005). Declarative pointing involves directing someone else's attention to an object for the purpose of sharing interest in that object, in contrast to imperative pointing, which involves directing someone else's attention to an object wanted by the pointer (something that other animals can do) (Liszkowski et al. 2004). Because declarative pointing is universally understood among cultures around the world (Tomasello et al. 2007), and no other animals do it, there must have been a strong selective pressure that favored the changes in the brain to enable us alone to perform this behavior.

Pointing is critically dependent on vision. Although deaf persons develop pointing, blind persons do not (Butterworth et al. 2002). By about 12 months of age, we can point declaratively (Liszkowski et al. 2004).

The development of declarative pointing is interesting in part because it indicates the earliest age at which we are beginning to understand triadic relationships involving ourselves, another person, and an object (Tomasello 2000; Liszkowski et al. 2004).

The main purpose of pointing is to shift someone else's attention to the visual periphery, and, secondarily, to identify objects of interest (Bangerter 2004; Liszkowski et al. 2004). When we point, we are most accurate at aiming to targets in the lower visual field. We also point faster to targets in the lower visual field (Danckert and Goodale 2001). Head, gaze, and pointing orientation all indicate direction of interest, and we process this information from others automatically (Driver et al. 1999; Langton and Bruce 2000; Langton et al. 2000). Automatic behaviors are often a matter of life and death, done because we either cannot afford the time to think about possible responses or we cannot afford to think about them and do other things at the same time.

Our response to someone else's pointing finger is to quickly orient our gaze in the same direction (Driver et al. 1999; Langton and Bruce 2000; Langton et al. 2000). We are much better at following the pointing of others to the visual periphery than to central locations (Butterworth and Itakura 2000) and we have faster reaction times when asked to respond to pictures of people looking and pointing down rather than up (Langton and Bruce 2000). Our automatic orientation to the gaze and pointing of others suggests that it is influenced by the SC-pulvinar visual system.

Tomasello and colleagues (2007) hypothesized that cooperative communication was the driving force behind the evolution of declarative pointing. They suggested that cooperative communication involves three classes of social intention or motive—the desire for others to (1) feel an emotion, (2) gain information, and (3) act. Expressed in this way, however, cooperative communication may not be limited to humans, e.g., many social animals have alarm calls that could be argued to involve those classes of social intention, and it still does not address why declarative pointing evolved only in hominins.

What was it outside central vision and in the lower visual field that was so urgent for our ancestors to see that it caused neurological changes to enable us to turn automatically in the direction of a gaze and a pointing finger? What in the lower visual field needed fast and accurate localization? In the absence of any other plausible evolutionary argument, I will freely speculate that venomous snakes, particularly those on the ground, could have acted as that selective pressure. I cannot think of any other object in the lower visual field that would have been more difficult to see and more unforgiving if missed. Of course, after perhaps several million years of declarative pointing, it is by now a useful tool for alerting others to additional objects. For example, the sense of urgency and emotion that declarative pointing can convey today is well illustrated by the famous photograph of three people simultaneously and independently gazing and pointing in the same direction to locate the shooter in the moments after Martin Luther King Jr. was assassinated.

Pointing as a Gesture Leading to Language

Gesturing in general has been hypothesized to have led to the evolution of language (e.g., Corballis 1992, 2003; Armstrong et al. 1994; Arbib 2005; Hopkins et al. 2005). If gesturing led to language, then, again, we have to explain why language did not also evolve in other primates that gesture, e.g., chimpanzees. Bipedalism is usually offered as the crucial difference but it is not clear how bipedalism per se would have operated as the selective pressure favoring language. If we limit gestures to declarative pointing, however, then the lack of language in other primates and its existence in humans is not so difficult to explain. Indeed, recent reviews are beginning to link pointing directly to language learning and creation (Goldin-Meadow 2007; Tomasello et al. 2007).

Pointing is a social activity; no one points when alone. Autism, a brain disorder that impairs social communication, affects children so that they are very much alone in a highly social environment. They have difficulty developing social relationships, they are poorly skilled in their use of language, and they do not point declaratively (Mundy et al. 1986; Baron-Cohen 1989, 1995). In 1995 being alone almost cost a young boy from the Pokot tribe his life when he was bitten on the right leg by a puff adder. He was in the bush herding his family's goats on the Laikipia Plateau in Kenya and did not see the snake in time. A neighbor eventually found him crying and sitting against a whistling thorn acacia. If it had been a cobra, the neighbor would have found him dead, but puff adder venom works differently. Like rattlesnake venom, it mainly destroys tissue, and there is often time to treat the bite. The boy was taken to the local infirmary after it was clear that the tobacco plug that was forced into the wound was not doing any good. They could not do anything for him at the infirmary, either, and that was when he was brought to our house. It was close to midnight by that time. While I stayed behind with our year-old son, my husband drove the boy and his mother an hour away to Nanyuki where he was admitted to the small but excellent Cottage Hospital. At one point during his stay, the boy's entire right side swelled up and gangrene threatened to set in. Saving the boy required slicing open his skin to relieve the pressure but that meant blood transfusions had to be given. The doctor was concerned that giving blood to the boy would increase his chances of acquiring HIV. Gangrene was a more immediate and certain outcome, however, and so blood transfusions were performed but kept to the minimum considered necessary. The doctor walked a fine line superbly. After three weeks in the hospital, the boy was able to return home with a story to tell for the rest of his life.

The treatment that saved the boy's life was not available for *Homo ergaster,* the first hominin fully committed to a bipedal life. A bite from a puff adder would have meant certain agonizing death for a young, lone boy 1.5 million years ago. Because more eyes are more effective at spotting danger, the chance of seeing a snake before stepping too near to it would have increased with others around. Declarative pointing undoubtedly would have reduced the frequency of deadly snakebites in the social groups of our hominin ancestors.

Furthermore, the message would have been easily understood by others when the first hominin pointed to a snake while vocalizing and looking fearfully at it. The three classes of social intention in declarative pointing would have all been present: wanting others to feel an emotion (fear), wanting them to gain information (the snake is there), and wanting them to do something (e.g., avoid walking near it). In this context, declarative pointing, as with many other evolutionary innovations, would have merely piggybacked onto an existing system because the other evolutionary precursors to human language were present long before declarative pointing and hominins arose. Non-human primates have a communication system with a formal structure that resembles the structure of spoken language, and it has largely come about as a result of living in a social environment (Cheney and Seyfarth 2005). In addition, some primate species have distinct vocalizations for snakes (Seyfarth et al. 1980), which are rarely produced except in social settings (Cheney and Seyfarth 2005).

You might argue that vocalizing and gazing should have been sufficient for alerting our hominin ancestors because these behaviors work well enough for other primates. I would counter that although vocalizing and gazing do work well, it is much easier for us to see snakes when pointing accompanies those behaviors and draws the eyes closer to the target, and the same would have applied to our bipedal hominin ancestors. Sometimes, even when the vervets alerted me to a snake by their vocalizations and intense gazing into the vegetation, I could not see it no matter what angle I tried. If the vervets could have pointed to the snake, I might have had greater success. Systematic studies confirm that pointing is more effective than just head and gaze orientation in directing attention to intended targets (Butterworth and Itakura 2000).

The connections between declarative pointing and language are numerous. First, pointing is gestural deixis, gestural emphasis designed to draw attention to certain things. Verbal deixis uses emphatic speech to emphasize certain words in a sentence, and it can often accompany or replace gestural deixis. Second, we typically reduce the number of words we speak and rely more on pointing when we want to direct someone's attention to a specific target closer to them (Bangerter 2004). These two findings illustrate that declarative pointing and language are interchangeable to some degree. Third, declarative pointing is a developmental precursor to language. Babies point before they speak, and 10–11-month-olds who look for longer and point at targets to which their mothers gaze acquire language at a faster rate by age two than those who look for shorter periods and do not point (Brooks and Meltzoff 2007) (Figure E.1). The responses

Figure E.1   Pointing at objects and longer gazing at those objects by 10-month-old infants are correlated with greater vocabulary at 30 months of age.

of eight-month-old babies to the direction of the mother's head, gaze, and pointing also predict individual differences in vocabulary development at 30 months (Morales et al. 2000; see also Tomasello et al. 2007). Fourth, declarative pointing and language both require an understanding of triadic relationships involving oneself, another, and a target (Degos 2001; Butterworth et al. 2002). Finally, recall that people with autism have poor social interactions with others, do not point declaratively, and have language deficits. Such people heartbreakingly reinforce the social nature of both pointing and language.

Huynh Sanh Thong, a MacArthur Fellow, has developed a linguistic argument that snakes were ultimately responsible for the origin of language because mothers needed to warn their children about them (Thong 1997). Declarative pointing might easily be incorporated into his argument. I offer that snakes gave bipedal hominins, who were already equipped with a non-human primate communication system, the evolutionary nudge to begin pointing to communicate for social good, a critical step toward the evolution of language, and all that followed to make us who we are today.

Neurological Terms and Some of Their Functions

Amygdala Responsive to auditory and visual stimuli; gaze direction; mediates expression of fearful behavior, including motor expression of fear, e.g., freezing; acquisition of, and learning about, fearful stimuli; acquisition of spatial memory; modulation of memory storage, especially of negatively arousing stimuli; involved in recognition of fearful faces.

Cuneiform nucleus Motor expressions of fear, e.g., freezing, darting, and running.

Dorsolateral prefrontal cortex (DLPFC) Saccades; attention; spatial localization of objects; memory of objects in nearby space; reaching; skilled hand movements.

Frontal eye field Involved in searching and scanning for salient stimuli; anticipatory activity before target appears; saccade-related activity.

Inferotemporal cortex (IT) Responsive to faces and other complex visual stimuli; object recognition.

Koniocellular (K) pathway Rapid responsiveness to motion; blue hues; luminance; auditory stimuli; attention and arousal; eye movements; and possibly facilitation of pop out.

Locus coeruleus (LC) Freezing behavior; primary central nervous system (CNS) source of norepinephrine, which is involved in increased attention, vigilance, learning, and enhanced memory, especially of negative experiences.

Magnocellular (M) pathway Responsive to motion and contrast; vision for motion detection.

Middle temporal cortex (MT) Involved in search of salient stimuli; motion; visual tracking.

Parvocellular (P) pathway Central vision; responsive to color and form; object perception; vision for object perception.

Periaqueductal gray (PAG) Freezing behavior.

Posterior parietal cortex (PPC) Saccades; attention; spatial localization of objects; memory of objects in nearby space; visually guided reaching and grasping.

Pulvinar, dorsal (medial and dorsal lateral pulvinar) Eye movements; gaze direction; orienting to salient stimuli; coordinates and integrates selective spatial attention; processing visual threats.

Pulvinar, inferior (inferior and ventral lateral pulvinar) Attention; selective visual processing; tuning out irrelevant stimuli; modulates activity in the V2 in anthropoids.

Superior colliculus (SC) Responsive to visual stimuli, especially luminance and movement; sends visual signals rapidly to deeper layers for orienting head and eyes; involved in visually guided reaching and grasping, and automatic visual detection of dangerous objects.

V1 (primary visual area) Responsive to simple visual stimuli, e.g., orientation and color; component perception.

V2 Responsive to simple and dual visual stimuli, e.g., color and orientation; color and disparity; stereopsis; global perception.

V4 Responsive to complex visual images, e.g., faces; attention to less physically prominent stimuli.

References

Adams, D. L., and Horton, J. C. 2006. Monocular cells without ocular dominance columns. *J. Neurophysiol.* 96: 2253–2264.

Adams, M. M., Hof, P. R., Gattass, R., Webster, M. J., and Ungerleider, L. G. 2000. Visual cortical projections and chemoarchitecture of macaque monkey pulvinar. *J. Comp. Neurol.* 419: 377–393.

Adolphs, R., Gosselin, F., Buchanan, T. W., Tranel, D., Schyns, P., and Damasio, A. R. 2005. A mechanism for impaired fear recognition after amygdala damage. *Nature* 433: 68–72.

Aggleton, J. P., and Saunders, R. C. 2000. The amygdala—what's happened in the last decade? In Aggleton, J. P., ed., *The Amygdala: A Functional Analysis.* New York: Oxford University Press, pp. 1–30.

Aiello, L. C., Bates, N., and Joffe, T. 2001. In defense of the expensive tissue hypothesis. In Falk, D., and Gibson, K. R., eds., *Evolutionary Anatomy of the Primate Cerebral Cortex.* Cambridge: Cambridge University Press, pp. 57–78.

Albino, A. M., and Montalvo, C. I. 2006. Snakes from the Cerro Azul Formation (Upper Miocene), Central Argentina, with a review of fossil viperids from South America. *J. Vert. Paleontol.* 26: 581–587.

Alexander, R. D. 1974. The evolution of social behavior. *Ann. Rev. Ecol. Syst.* 5: 324–382.

Ali, J. R., and Aitchison, J. C. 2008. Gondwana to Asia: plate tectonics, paleogeography and the biological connectivity of the Indian sub-continent from the Middle Jurassic through latest Eocene (166–35 Ma). *Earth Sci. Rev.* 88: 145–166.

Allman, J. M. 1977. Evolution of the visual system in the early primates. In Sprague, J. M., and Epstein, A. N., eds., *Progress in Psychobiology, Physiology and Psychology,* vol. 7. New York: Academic Press, pp. 1–53.

Allman, J. M. 1999. *Evolving Brains.* New York: Scientific American Library.

Allman, J. M., and McGuinness, E. 1988. Visual cortex in primates. In Steklis, H. D., and Erwin, J., eds., *Comparative Primate Biology,* vol. 4. New York: Alan R. Liss, pp. 279–326.

Allman, J. M., and Zucker, S. 1990. Cytochrome oxidase and functional coding in primate striate cortex: a hypothesis. *Cold Spring Harb. Symp. Quant. Biol.* 55: 979–982.

Amaral, D. G. 2002. The primate amygdala and the neurobiology of social behavior: implications for understanding social anxiety. *Biol. Psychiat.* 51: 11–17.

Amaral, D. G. 2003. The amygdala, social behavior, and danger detection. *Ann. New York Acad. Sci.* 1000: 337–347.

Amaral, D. G., and Price, J. L. 1984. Amygdalo-cortical projections in the monkey *(Macaca fascicularis). J. Comp. Neurol.* 230: 465–496.

Amaral, D. G., Price, J. L., Pitkänen, A., and Carmichael, S. T. 1992. Anatomical organization of the primate amygdaloid complex. In Aggleton, J. P., ed., *The Amygdala: Neurobiological Aspects of Emotion, Memory, and Mental Dysfunction.* New York: Wiley-Liss, pp. 1–66.

Amrine-Madsen, H., Hoepfli, K. P., Wayne, R. K., and Springer, M. S. 2003. A new phylogenetic marker, apolipoprotein B, provides compelling evidence for eutherian relationships. *Mol. Phylogen. Evol.* 28: 225–240.

Andersen, P. R., Barbacid, M., and Tronick, S. R. 1979. Evolutionary relatedness of viper and primate endogenous viruses. *Science* 204: 318–321.

Apesteguía, S., and Zaher, H. 2006. A Cretaceous terrestrial snake with robust hindlimbs and a sacrum. *Nature* 440: 1037–1040.

Arbib, M. A. 2005. From monkey-like action recognition to human language: an evolutionary framework for neurolinguistics. *Behav. Brain Sci.* 28: 105–167.

Arendes, L. 1994. Superior colliculus activity related to attention and to connotative stimulus meaning. *Cog. Brain Res.* 2: 65–69.

Armstrong, D. F., Stokoe, W. C., and Wilcox, S. E. 1994. Signs of the origin of syntax. *Curr. Anthropol.* 35: 349–368.

Armstrong, E. 1983. Relative brain size and metabolism in mammals. *Science* 220: 1302–1304.

Arnason, U., Gullberg, A., and Berguete, A. S. 1998. Molecular timing of primate divergences as estimated by two nonprimate calibration points. *J. Mol. Evol.* 47: 718–727.

Asher, R. J., Meng, J., Wible, J. R., McKenna, M. C., Rougier, G. W., Dashzeveg, D., and Novacek, M. J. 2005. Stem Lagomorpha and the antiquity of Glires. *Science* 307: 1091–1094.

Aston-Jones, G., Chiang, C., and Alexinsky, T. 1991. Discharge of noradrenergic locus coeruleus neurons in behaving rats and monkeys suggests a role in vigilance. *Prog. Brain Res.* 88: 501–520.

Aston-Jones, G., Rajkowski, J., and Kubiak, P. 1997. Conditioned responses of monkey locus coeruleus neurons anticipate acquisition of discriminative behavior in a vigilance task. *Neuroscience* 80: 697–715.

Aston-Jones, G., Rajkowski, J., Kubiak, P., and Alexinsky, T. 1994. Locus coeruleus neurons in monkey are selectively activated by attended cues in a vigilance task. *J. Neurosci.* 14: 4467–4480.

Baleydier, C., and Mauguiere, F. 1985. Anatomical evidence for medial pulvinar connections with the posterior cingulate cortex, the retrosplenial area, and the posterior parahippocampal gyrus in monkeys. *J. Comp. Neurol.* 232: 219–228.

Baleydier, C., and Mauguiere, F. 1987. Network organization of the connectivity between parietal area 7 posterior cingulate cortex and medial pulvinar nucleus: a double fluorescent tracer study in monkey. *Exp. Brain Res.* 66: 385–393.

Bangerter, A. 2004. Using pointing and describing to achieve joint focus of attention in dialogue. *Psychol. Sci.* 15: 415–419.

Barbur, J. L., Ruddock, K. H., and Waterfield, V. A. 1980. Human visual responses in the absence of the geniculo-calcarine projection. *Brain* 103: 905–928.

Barbur, J. L., Weiskrantz, L., and Harlow, J. A. 1999. The unseen color aftereffect of an unseen stimulus: insight from blindsight into mechanism of color afterimages. *Proc. Natl. Acad. Sci.* 96: 11637–11641.

Baron, G., Stephan, H., and Frahm, H. D. 1996. *Comparative Neurobiology in Chiroptera*, vol. 3. Basel: Birkhauser Verlag.

Baron-Cohen, S. 1989. Perceptual role taking and protodeclarative pointing in autism. *Brit. J. Develop. Psychol.* 7: 113–127.

Baron-Cohen, S. 1995. *Mindblindness: An Essay on Autism and Theory of Mind.* Cambridge, MA: MIT Press.

Barrett, L., Gaynor, D., Rendall, D., Mitchell, D., and Henzi, S. P. 2004. Habitual cave use and thermoregulation in chacma baboons *(Papio hamadryas ursinus)*. *J. Hum. Evol.* 46: 215–222.

Barton, R. A. 1996. Neocortex size and behavioural ecology in primates. *Proc. R. Soc. Lond. B* 263: 173–177.

Barton, R. A. 1998. Visual specialization and brain evolution in primates. *Proc. R. Soc. Lond. B* 265: 1933–1937.

Barton, R. A. 1999. The evolutionary ecology of the primate brain. In Lee, P. C., ed., *Comparative Primate Socioecology.* Cambridge: Cambridge University Press, pp. 167–203.

Barton, R. A. 2000. Primate brain evolution: cognitive demands of foraging or of social life? In Boinski, S., and Garber, P. A., eds., *On the Move: How and Why Animals Travel in Groups.* Chicago: University of Chicago Press, pp. 204–237.

Barton, R. A. 2004. Binocularity and brain evolution in primates. *Proc. Natl. Acad. Sci.* 101: 10113–10115.

Barton, R. A., and Aggleton, J. P. 2000. Primate evolution and the amygdala. In Aggleton, J. P., ed., *The Amygdala: A Functional Analysis.* New York: Oxford University Press, pp. 479–508.

Barton, R. A., Aggleton, J. P., and Grenyer, R. 2003. Evolutionary coherence of the mammalian amygdala. *Proc. R. Soc. Lond. B* 270: 539–543.

Barton, R. A., Purvis, A., and Harvey, P. H. 1995. Evolutionary radiation of visual and olfactory brain systems in primates, bats and insectivores. *Phil. Trans. R. Soc. Lond. B* 348: 381–932.

Beard, K. C. 2002. Basal anthropoids. In Hartwig, W. C., ed., *The Primate Fossil Record*. Cambridge: Cambridge University Press, pp. 133–149.

Beard, K. C. 2004. *The Hunt for the Dawn Monkey*. Berkeley: University of California Press.

Beck, R. A., Burbank, D. W., Sercombe, W. J., Riley, G. W., Barndt, J. K., Berry, J. R., Afzal, J., Khan, A. M., Jergen, H., Metje, J., Cheema, A., Shafique, N. A., Lawrence, R. D., and Khan, M. A. 1995. Stratigraphic evidence for an early collision between northwest India and Asia. *Nature* 373: 55–58.

Bender, D. B., and Butter, C. M. 1987. Comparison of the effects of superior colliculus and pulvinar lesions on visual search and tachistoscopic pattern discrimination in monkeys. *Exp. Brain Res.* 69: 140–154.

Bender, D. B., and Youakim, M. 2001. Effect of attentive fixation in macaque thalamus and cortex. *J. Neurophysiol.* 85: 219–234.

Benevento, L. A., and Port, J. D. 1995. Single neurons with both form/color differential responses and saccade-related responses in the nonretinotopic pulvinar in the behaving macaque monkey. *Vis. Neurosci.* 12: 523–544.

Benevento, L. A., and Rezak, M. 1976. The cortical projections of the inferior pulvinar and adjacent lateral pulvinar in the rhesus monkey *(Macaca mulatta):* an autoradiographic study. *Brain Res.* 108: 1–24.

Berridge, C. W., and Waterhouse, B. D. 2003. The locus coeruleus-noradrenergic system: modulation of behavioral state and state-dependent cognitive processes. *Brain Res. Rev.* 42: 33–84.

Bininda-Emonds, O. R. P., Cardillo, M., Jones, K. E., MacPhee, R. D. E., Beck, R. M. D., Grenyer, R., Price, S. A., Vos, R. A., Gittleman, J. L., and Purvis, A. 2007. The delayed rise of present-day mammals. *Nature* 446: 507–512.

Bishop, A. 1964. Use of the hand in lower primates. In Buettner-Janusch, J., ed., *Evolutionary and Genetic Biology of Primates,* vol. 2. New York: Academic Press, pp. 133–225.

Bloch, J. I., and Boyer, D. M. 2002. Grasping primate origins. *Science* 298: 1606–1610.

Bloch, J. I., Silcox, M. T., Boyer, D. M., and Sargis, E. J. 2007. New Paleocene skeletons and the relationship of plesiadapiforms to crown-clade primates. *Proc. Natl. Acad. Sci.* 104: 1159–1164.

Blythe, I. M., Kennard, C., and Ruddock, K. H. 1987. Residual vision in patients with retrogeniculate lesions of the visual pathways. *Brain* 110: 887–905.

Boesch, C. 1991. The effects of leopard predation on grouping patterns in forest chimpanzees. *Behaviour* 117: 220–236.

Boinski, S. 1988. Use of a club by a wild white-faced capuchin *(Cebus capucinus)* to attack a venomous snake *(Bothrops asper). Amer. J. Primatol.* 14: 177–179.

Bossuyt, F., and Milinkovitch, M. 2001. Amphibians as indicators of early Tertiary "out-of-India" dispersal of vertebrates. *Science* 292: 93–95.

Brandão, M. L., Troncoso, A. C., de Souza Silva, M. A., and Huston, J. P. 2003. The relevance of neuronal substrates of defense in the midbrain tectum to anxiety and stress: empirical and conceptual considerations. *Europ. J. Pharmacol.* 463: 225–233.

Briggs, J. C. 2003. The biogeographical and tectonic history of India. *J. Biogeogr.* 30: 381–388.

Brooke, M. de L., Hanley, S., and Laughlin, S. B. 1999. The scaling of eye size with body mass in birds. *Proc. R. Soc. Lond. B* 266: 405–412.

Brooks, R., and Meltzoff, A. N. 2007. Infant gaze following and pointing predict accelerated vocabulary growth through two years of age: a longitudinal, growth curve modeling study. *J. Child Lang.* 35: 207–220.

Brothers, L., Rign, B., and Kling, A. 1990. Response of neurons in the macaque amygdala to complex social stimuli. *Behav. Brain Res.* 41: 199–213.

Brown, G. C., and Bal-Price, A. 2003. Inflammatory neurodegeneration mediated by nitric oxide, glutamate, and mitochondria. *Mol. Neurobiol.* 27: 325–355.

Brown, L. 1976. *Eagles of the World.* New York: Universe Books.

Buchanan-Smith, H. M., Smith, A. C., Surridge, A. K., Prescott, M. J., Osorio, D., and Mundy, N. I. 2005. The effect of sex and color vision status on prey capture by captive and wild tamarins (*Saguinus* spp.). *Amer. J. Primatol.* 66: 49.

Bullier, J., Girard, P., and Salin, P.-A. 1994. The role of area 17 in the transfer of information to extrastriate visual cortex. In Peters, A., and Rockland, K. S., eds., *Cerebral Cortex,* vol. 10. New York: Plenum, pp. 301–330.

Bullier, J., and Kennedy, H. 1983. Projection of the lateral geniculate nucleus onto cortical area V2 in the macaque monkey. *Exp. Brain Res.* 53: 168–172.

Burney, D. A. 2002. Sifaka predation by a large boa. *Folia Primatol.* 73: 144–145.

Busse, C. 1980. Leopard and lion predation upon chacma baboons living in the Moremi Wildlife Reserve. *Botswana Notes Rec.* 12: 15–21.

Butler, A. B. 1994. The evolution of the dorsal thalamus of jawed vertebrates, including mammals: cladistic analysis and a new hypothesis. *Brain Res. Rev.* 19: 29–65.

Butterworth, G., Franco, F., McKenzie, B., Graupner, L., and Todd, B. 2002. Dynamic aspects of visual event perception and the production of pointing by human infants. *J. Develop. Psychol.* 20: 1–24.

Butterworth, G., and Itakura, S. 2000. How the eyes, head and hand serve definite reference. *Brit. J. Develop. Psychol.* 18: 25–50.

Byrne, R., and Whiten, A., eds. 1988. *Machiavellian Intelligence.* Oxford: Oxford University Press.

Cadle, J. E. 1987. Geographic distribution: problems in phylogeny and zoogeography. In Seigel, R. A., Collins, J. T., and Novak, S. S., eds., *Snakes: Ecology and Evolutionary Biology.* New York: Macmillan, pp. 77–105.

Cadle, J. E. 1988. Phylogenetic relationships among advanced anakes. *U. Calif. Publ. Zool.* 119: 1–77.

Cahill, L., and McGaugh, J. L. 1998. Mechanisms of emotional arousal and lasting declarative memory. *Trends Neurosci.* 21: 294–299.

Calkins, D. J. 2001. Seeing with S cones. *Prog. Retin. Eye Res.* 20: 255–287.

Cantalupo, C., McCain, D., and Ward, J. P. 2002. Functions of head-cocking in Garnett's greater bush baby (*Otolemur garnettii*). *Int. J. Primatol.* 23: 203–221.

Capaldi, R. A. 1990. Structure and function of cytochrome c oxidase. *Ann. Rev. Biochem.* 59: 569–596.

Caprette, C. L., Lee, M. S. Y., Shine R., Mokany A., and Downhower, J. F. 2004. The origin of snakes (Serpentes) as seen through eye anatomy. *Biol. J. Linn. Soc.* 81: 469–482.

Carey, D. P., Coleman, R. J., and Della Sala, S. 1997. Magnetic misreaching. *Cortex* 33: 639–652.

Carthew, S. M., and Goldingay, R. L. 1997. Non-flying mammals as pollinators. *Trends Evol. Ecol.* 12: 104–108.

Cartmill, M. 1972. Arboreal adaptations and the origin of the order Primates. In Tuttle, R., ed., *The Functional and Evolutionary Biology of Primates.* Chicago: Aldine-Atherton, pp. 97–122.

Cartmill, M. 1974. Rethinking primate origins. *Science* 184: 436–443.

Cartmill, M. 1992. New views on primate origins. *Evol. Anthropol.* 6: 105–111.

Casagrande, V. A. 1994. A third parallel visual pathway to primate area V1. *Trends Neurosci.* 17: 305–310.

Casagrande, V. A. 1999. The mystery of the visual system K pathway. *J. Physiol. (Lond.)* 517: 630.

Casagrande, V. A., and Royal, D. W. 2003. Parallel visual pathways in a dynamic system. In Kaas, J. H., and Collins, C. E., eds., *Primate Vision.* Boca Raton, FL: CRC Press, pp. 1–28.

Casagrande, V. A., and Xu, X. 2003. Parallel visual pathways: a comparative perspective. In Chalupa, L., and Werner, J. S., eds. *The Visual Neurosciences.* Cambridge, MA: MIT Press, pp. 494–506.

Casagrande, V. A., Yazar, F., Jones, K. D., and Ding, Y. 2007. The morphology of the koniocellular axon pathway in the macaque monkey. *Cerebral Cortex* 17: 2334–2345.

Casanova, C., Merabet, L., Desautels, A., and Minville, K. 2001. Higher-order motion processing in the pulvinar. *Prog. Brain Res.* 134: 71–82.

Case, J. A. 2002. A new biogeographic model for dispersal of Late Cretaceous vertebrates into Madagascar and India. *J. Vertebrate Paleontol.* 22: 42A.

Cave, A. J. E. 1973. The primate nasal fossa. *Biol. J. Linn. Soc.* 5: 377–387.

Cenci, M. A., Whishaw, I. Q., and Schallert, T. 2002. Animal models of neurological deficits: how relevant is the rat? *Nature Rev. Neurosci.* 3: 574–579.

Chalfin, B. P., Cheung, D. T., Muniz, J. A. P. C., de Lima Silveira, L. C., and Finlay, B. L. 2007. Scaling of neuron number and volume of the pulvinar complex in New World primates: comparisons with humans, other primates, and mammals. *J. Comp. Neurol.* 504: 265–274.

Chalupa, L. M. 1991. Visual function of the pulvinar. In Leventhal, A. G., ed., *The Neural Basis of Visual Function.* Boca Raton, FL: CRC Press, pp. 140–159.

Chao, L. L., and Martin, A. 2000. Representation of manipulable man-made objects in the dorsal stream. *Neuroimage* 12: 478–484.

Chapman, C. A. 1986. *Boa constrictor* predation and group response in white-faced cebus monkeys. *Biotropica* 18: 171–172.

Charles-Dominique, P. 1977. *Ecology and Behaviour of Nocturnal Prosimians.* New York: Columbia University Press.

Chatterjee, S., and Callaway, E. M. 2003. Parallel colour-opponent pathways to primary visual cortex. *Nature* 426: 668–671.

Chaves, R., Sampaio, I., Schneider, M. P., Schneider, H., Page, S. L., and Goodman, M. 1999. The place of *Callimico goeldii* in the Callitrichine phylogenetic tree: evidence from von Willebrand factor gene intron II sequences. *Mol. Phylogen. Evol.* 13: 392–404.

Cheney, D. L., and Seyfarth, R. M. 2005. Constraints and preadaptation in the earliest stages of language evolution. *Linguistic Rev.* 22: 135–159.

Cheney, D. L., and Wrangham, R. W. 1987. Predation. In Smuts, B. B., Cheney, D. L., Seyfarth, R. M., Wrangham, R. W., and Struhsaker, T. T., eds., *Primate Societies.* Chicago: University of Chicago Press, pp. 227–239.

Chism, J., Rowell, T. E., and Olson, D. K. 1984. Life history patterns of female patas monkeys. In Small, M. D., ed., *Female Primates: Studies by Women Primatologists.* New York: Alan R. Liss, pp. 175–190.

Choi, D. D. 1988. Glutamate neurotoxicity and diseases of the nervous system. *Neuron* 1: 623–634.

Ciochon, R. L., and Gunnell, G. F. 2002. Eocene primates from Myanmar: historical perspectives on the origin of Anthropoidea. *Evol. Anthropol.* 11: 156–166.

Clarke, J. A., Tambussi, C. P., Noriega, J. I., Erickson, G. M., and Ketcham, R. A. 2005. Definitive fossil evidence for the extant avian radiation in the Cretaceous. *Nature* 433: 305–308.

Clower, D. M., West, R. A., Lynch, J. C., and Strick, P. L. 2001. The inferior parietal lobule is the target of output from the superior colliculus, hippocampus, and cerebellum. *J. Neurosci.* 21: 6283–6291.

Clutton-Brock, T. H., and Harvey, P. H. 1980. Primates, brains and ecology. *J. Zool., Lond.* 190: 309–323.

Coates, A. G., and Obando, J. A. 1996. The geologic evolution of the Central American isthmus. In Jackson, J. B. C., Budd, A. F., and Coates, A. G., eds., *Evolution and Environment in Tropical America.* Chicago: University of Chicago Press, pp. 21–56.

Collins, C. E., Stepniewska, I., and Kaas, J. H. 2001. Topographic patterns of V2 cortical connections in a prosimian primate *(Galago garnetti). J. Comp. Neurol.* 431: 155–167.

Colquhoun, I. C. 2006. Predation and cathermerality. *Folia Primatol.* 77: 143–165.

Comoli, E., Coizet, V., Boyes, J., Bolam, J. P., Canteras, N. S., Quirk, R. H., Overton, P. G., and Redgrave, P. 2003. A direct projection from superior colliculus to substantia nigra for detecting salient visual events. *Nature Neurosci.* 6: 974–980.

Condo, G. J., and Casagrande, V. A. 1990. Organization of cytochrome oxidase staining in the visual cortex of nocturnal primates (*Galago crassicaudatus* and *Galago senegalensis*): I. adult patterns. *J. Comp. Neurol.* 293: 632–645.

Conti, E., Eriksson, T., Schoneneberger, J., Systsma, K. J., and Baum, D. A. 2002. Early Tertiary out-of-India dispersal of Crypteroniaceae. *Evolution* 56: 1931–1942.

Cook, M., and Mineka, S. 1989. Observational conditioning of fear to fear-relevant versus fear-irrelevant stimuli in rhesus monkeys. *J. Abnormal Psychol.* 98: 448–459.

Cooper, A., Lalueza-Fox, C., Anderson, S., Rambaut, A., Austin, J., and Ward, D. 2001. Complete mitochondrial genome sequences of two extinct moas clarify ratite evolution. *Nature* 409: 704–707.

Cooper, A., and Penny, D. 1997. Mass survival of birds across the Cretaceous-Tertiary boundary: molecular evidence. *Science* 275: 1109–1113.

Cooper, H. M., Herbin, M., and Nevo, E. 1993. Visual system of a naturally microphthalmic mammal: the blind mole rat, *Spalax ehrenbergi. J. Comp. Neurol.* 328: 313–350.

Corballis, M. G. 1992. On the evolution of language and generativity. *Cognition* 44: 197–226.

Corballis, M. G. 2003. From mouth to hand: gesture, speech, and the evolution of right-handedness. *Behav. Brain Sci.* 26: 199–260.

Corréa, H. K. M., and Coutinho, P. E. G. 1997. Fatal attack of a pit viper, *Bothrops jararaca,* on an infant buffy-tufted ear marmoset *(Callithrix aurita). Primates* 38: 215–217.

Coss, R. G. 1991. Context and animal behavior III: the relationship between early development and evolutionary persistence of ground squirrel antisnake behavior. *Ecol. Psychol.* 3: 277–315.

Coss, R. G. 2003. The role of evolved perceptual biases in art and design. In Voland, E., and Grammer, K., eds., *Evolutionary Aesthetics.* New York: Springer, pp. 69–130.

Coss, R. G., Guse, K. L., Poran, N. S., and Smith, D. G. 1993. Development of antisnake defenses in California ground squirrels *(Spermophilus beecheyi)*— 2. microevolutionary effects of relaxed selection from rattlesnakes. *Behaviour* 124: 137–164.

Coss, R. G., and Owings, D. H. 1985. Restraints on ground squirrel anti-predator behavior: adjustments over multiple time scales. In Johnson, T. D., and Pietrewicz, A. T., eds., *Issues in the Ecological Study of Learning.* London: Lawrence Erlbaum Associates, pp. 167–200.

Courjon, J.-H., Olivier, E., and Pélisson, D. 2004. Direct evidence for the contribution of the superior colliculus in the control of visually guided reaching movements in the cat. *J. Physiol.* 556: 675–681.

Covert, H. H. 2002. The earliest fossil primates and the evolution of prosimians: introduction. In Hartwig, W. C., ed., *The Primate Fossil Record.* Cambridge: Cambridge University Press, pp. 13–20.

Cowey, A., and Stoerig, P. 1989. Projection patterns of surviving neurons in the dorsal lateral geniculate nucleus following discrete lesions of striate cortex: implications for residual vision. *Exp. Brain Res.* 75: 631–638.

Cowey, A., and Stoerig, P. 1997. Visual detection in monkeys with blindsight. *Neuropsychology* 35: 929–939.

Cowey, A., Stoerig, P., and Bannister, M. 1994. Retinal ganglion cells labelled from the pulvinar nucleus in macaque monkeys. *Neuroscience* 61: 691–705.

Cracraft, J. 2001. Avian evolution, Gondwana biogeography and the Cretaceous-Tertiary mass extinction event. *Proc. R. Soc. Lond. B* 268: 459–469.

Cresho, H. S., Rasco, L. M., Rose, G. H., and Condo, G. J. 1992. Blob-like pattern of cytochrome oxidase staining in ferret visual cortex. *Soc. Neurosci. Abst.* 18: 298.

Crish, S. D., Dengler-Crish, C. M., and Catania, K. C. 2006. Central visual system of the naked mole-rat *(Heterocephalus glaber)*. *Anat. Rec. A* 288A: 205–212.

Crockett, C. M., and Eisenberg, J. F. 1987. Howlers: variations in group size and demography. In Smuts, B. B., Cheney, D. L., Seyfarth, R. M., Wrangham, R. W., and Struhsaker, T. T., eds., *Primate Societies*. Chicago: University of Chicago Press, pp. 54–68.

Crompton, R. H. 1995. "Visual predation," habitat structure, and the ancestral primate niche. In Alterman, L., Doyle, G., and Izard, M. K. eds., *Creatures of the Dark: The Nocturnal Prosimians*. New York: Plenum Press, pp. 11–30.

Cropp, S., and Boinski, S. 2000. The Central American squirrel monkey *(Saimiri oerstedii)*: introduced hybrid or endemic species? *Mol. Phylogen. Evol.* 16: 350–365.

Crother, B. I., Campbell, J. A., and Hillis, D. M. 1992. Phylogeny and historical biogeography of the palm-pitvipers, genus *Bothriechis:* biochemical and morphological evidence. In Campbell, J. A., and Brodie Jr., E. D., eds., *Biology of the Pitvipers*. Tyler, TX: Selva, pp. 1–19.

Curcio, C. A., and Harting, J. K. 1978. Organization of pulvinar afferents to area 18 in the squirrel monkey: evidence for stripes. *Brain Res.* 143: 155–161.

Dagosto, M. 1988. Implications of postcranial evidence for the origin of euprimates. *J. Hum. Evol.* 17: 35–56.

Dagosto, M. 2002. The origin and diversification of anthropoid primates. In Hartwig, W. C., ed., *The Primate Fossil Record*. Cambridge: Cambridge University Press, pp. 125–132.

Daltry, J. C., Wüster, W., and Thorpe, R. S. 1996. Diet and snake venom evolution. *Nature* 379: 537–540.

D'Amour, D. E., Hohmann, G., and Fruth, B. 2006. Evidence of leopard predation on bonobos. *Folia Primatol.* 77: 212–217.

Danckert, J., and Goodale, M. A. 2001. Superior performance for visually guided pointing in the lower visual field. *Exp. Brain Res.* 137: 303–308.

Darwin, C. R. 1872. *The Expression of the Emotions in Man and Animals*. London: John Murray.

Davis, M. 2000. The role of the amygdala in conditioned and unconditioned fear and anxiety. In Aggleton, J. P., ed., *The Amygdala: A Functional Analysis*. New York: Oxford University Press, pp. 213–287.

Dean, P., Redgrave, P., and Westby, G. W. M. 1989. Event or emergency? Two response systems in the mammalian superior colliculus. *Trends Neurosci.* 12: 137–147.

de Gelder, B., and Hadjikhani, N. 2006. Non-conscious recognition of emotional body language. *NeuroReport* 17: 583–586.

Degos, J.-D. 2001. Troubles de la désignation. *Rev. Neuropsychol.* 11: 257–265.

Dell'Omo, G., and Alleva, E. 1994. Snake odor alters behavior, but not pain sensitivity in mice. *Physiol. Behav.* 55: 125–128.

del Rayo Sánchez-Carbente, M. D., and Massieu, L. 1999. Transient inhibition of glutamate uptake in vivo induces neurodegeneration when energy metabolism is impaired. *J. Neurochem.* 72: 129–138.

Delsuc, F., Scally, M., Madsen, O., Stanhope, M. J., de Jong, W. W., Catzeflis, F. M., Springer, M. S., and Douzery, E. J. P. 2002. Molecular phylogeny of living xenarthrans and the impact of character and taxon sampling on the placental tree rooting. *Mol. Biol. Evol.* 19: 1656–1671.

Deppe, A. M. 2005. Visual predator recognition and response in wild mouse lemurs *(Microcebus rufus)* in Ranomafana National Park, Madagascar. *Amer. J. Primatol.* 66: 97–98.

Deppe, A. M. 2006. Snake recognition in wild brown mouse lemurs *(Microcebus rufus)*. *Amer. J. Primatol.* 68: 34.

De Valois, R. L., and Jacobs, G. H. 1968. Primate color vision. *Science* 162: 533–540.

Dewey, J. F., Cande, S., and Pitman III, W. C. 1989. Tectonic evolution of the India/Eurasia collision zone. *Eclogae Geol. Helv.* 82: 717–734.

de Winter, W., and Oxnard, C. E. 2001. Evolutionary radiations and convergences in the structural organization of mammalian brains. *Nature* 409: 710–714.

DeYoe, E. A., and Van Essen, D. C. 1985. Segregation of efferent connections and receptive field properties in visual area V2 of the macaque. *Nature* 317: 58–61.

DeYoe, E. A, and Van Essen, D. C. 1988. Concurrent processing streams in monkey visual cortex. *Trends Neurosci.* 11: 219–226.

Di Fiore, A. 2002. Predator sensitive foraging in ateline primates. In Miller, L. E., ed., *Eat or Be Eaten: Predator Sensitive Foraging Among Primates.* Cambridge: Cambridge University Press, pp. 242–267.

Di Fiore, A., and Campbell, C. J. 2007. The atelines: variation in ecology, behavior, and social organization. In Campbell, C. J., Fuentes, A., MacKinnon, K. C., Panger, M., and Bearder, S. K., eds., *Primates in Perspective.* New York: Oxford University Press, pp. 155–185.

Digweed, S. M., Fedigan, L. M., and Rendall, D. 2005. Variable specificity in the anti-predator vocalizations and behaviour of the white-faced capuchin, *Cebus capucinus. Behaviour* 142: 997–1021.

Dolan, R. J., and Vuilleumier, P. 2003. Amygdala automaticity in emotional processing. *Ann. New York Acad. Sci.* 985: 348–355.

Dominy, N. J., and Lucas, P. W. 2001. Ecological importance of trichromatic vision to primates. *Nature* 410: 363–366.

Douady, C. J., Chatelier, P. I., Madsen, O., de Jong, W. W., Catzeflis, F., Springer, M. S., and Stanhope, M. J. 2002. Molecular phylogenetic evidence confirming the Eulipotyphla concept and in support of hedgehogs as the sister group to shrews. *Mol. Phylogen. Evol.* 25: 200–209.

Doubell, T. P., Skaliora, I., Baron, J., and King, A. J. 2003. Functional connectivity between the superficial and deeper layers of the superior colliculus: an anatomical substrate for sensorimotor integration. *J. Neurosci.* 23: 6596–6607.

Dowling, H. G., Hass, C. A., Hedges, S. B., and Highton, R. 1996. Snake relationships revealed by slow-evolving proteins: a preliminary survey. *J. Zool., Lond.* 240: 1–28.

Doyle, A. C. 1890. *The Sign of Four*. London: Spencer Blackett.

Driver, J., Davis, G., Ricciardelli, P., Kidd, P., Mazwell, E., and Baron-Cohen, S. 1999. Gaze perception triggers automatic visuospatial orienting in adults. *Visual Cognit.* 6: 509–540.

Dunbar, R. I. M. 1988. *Primate Social Systems*. Ithaca, NY: Cornell University Press.

Dunbar, R. I. M. 1992. Neocortex size as a constraint on group size in primates. *J. Hum. Evol.* 20: 469–493.

Dunbar, R. I. M. 1998. The social brain hypothesis. *Evol. Anthropol.* 6: 178–190.

Dunbar, R. I. M. 2003. The social brain: mind, language, and society in evolutionary perspective. *Ann. Rev. Anthropol.* 32: 163–181.

Edelstam, C. 2001. Raptor vision, hearing and olfaction. In Ferguson-Lees, J., and Christie, D. A., eds., *Raptors of the World*. London: Christopher Helm, pp. 54–56.

Egi, N., Takai, M., Shigehara, N., and Tsubamoto, T. 2004. Body mass estimates for Eocene eosimiid and amphipithecid primates using prosimian and anthropoid scaling models. *Int. J. Primatol.* 25: 211–236.

Eizirik, E., Murphy, W. J., and O'Brien, S. J. 2001. Molecular dating and biogeography of the early placental mammal radiation. *J. Heredity* 92: 212–219.

Eizirik, E., Murphy, W. J., Springer, M. S., and O'Brien, S. J. 2004. Molecular phylogeny and dating of early primate divergences. In Ross, C., and Kay, R. F., eds., *Anthropoid Origins: New Visions*. New York: Kluwer Academic/Plenum Press, pp. 45–64.

Ellard, C. G., and Goodale, M. A. 1988. A functional analysis of the collicular output pathways: a dissociation of deficits following lesions of the dorsal tegmental decussation and the ipsilateral collicular efferent bundle in the Mongolian gerbil. *Exp. Brain Res.* 71: 307–319.

Elliot Smith, G. 1912. *The Evolution of Man*. Smithsonian Institution Annual Report. Washington, DC: Smithsonian Institution.

Emery, N. J. 2000. The eyes have it: the neuroethology, function and evolution of social gaze. *Neurosci. Biobehav. Rev.* 24: 581–604.

Ericson, P. G. P., Anderson, C. L., Britton, T., Elzanowski, A., Johansson, U. S., Källersjö, M., Ohlson, J. I., Parsons, T. J., Zuccon, D., and Mayr, G. 2006.

Diversification of Neoaves: integration of molecular sequence data and fossils. *Biol. Lett.* 2: 543–547.

Evans, S. E., Jones, M. E. H., and Krause, D. W. 2008. A giant frog with South American affinities from the late Cretaceous of Madagascar. *Proc. Nat. Acad. Sci.* 105: 2951–2956.

Fawcett, P. H. 1953. *Exploration Fawcett.* London: Hutchinson & Co., Ltd.

Feduccia, A. 1995. "Big bang" for tertiary birds? *Trends Ecol. Evol.* 18: 172–176.

Feldman, R. S., Meyer, J. S., and Quenzer, L. F. 1997. *Principles of Neuropsychoparmacology.* Sunderland, MA: Sinauer Associates.

Ferguson-Lees, J., and Christie, D. A. 2001. *Raptors of the World.* London: Christopher Helm.

Fernandez-Duque, E. 2003. Influences of moonlight, ambient temperature, and food availability on the diurnal and nocturnal activity of owl monkeys *(Aotus azarai). Behav. Ecol. Sociobiol.* 54: 431–440.

Fielding, J., Georgiou-Karistianis, N., and White, O. 2006. The role of the basal ganglia in the control of automatic visuospatial attention. *J. Inter. Neuropsychol. Soc.* 12: 657–667.

Fleagle, J. G. 1999. *Primate Adaptation and Evolution,* 2nd ed. New York: Academic Press.

Florence, S. L., and Kaas, J. H. 1992. Ocular dominance columns in area 17 of Old World macaque and talapoin monkeys: complete reconstructions and quantitative analysis. *Vis. Neurosci.* 8: 449–462.

Foerster, S. 2008. Two incidents of snake attack on juvenile blue and Sykes monkeys (*Cercopithecus mitis stuhmanni* and *C. m. albogularis*). *Primates* 49: 300–303.

Folie, A. E., and Codrea, V. 2005. New lissamphibians and squamates from the Maastrichtian of Hateg Basin, Romania. *Acta Palaeontol. Polonica* 50: 57–71.

Foote, S. L., Berridge, C. W., Adams, L. M., and Pineda, J. A. 1991. Electrophysiological evidence for the involvement of the locus coeruleus in alerting, orienting, and attending. *Prog. Brain Res.* 88: 521–532.

Fredrikson, M., Wik, G., Annas, P., Ericson, K., and Stone-Elander, S. 1995. Functional neuroanatomy of visually elicited simple phobic fear: additional data and theoretical analysis. *Psychophysiology* 32: 43–48.

Frey, F. A., Coffin, M. F., Wallace, P. J., and Weis, D. 2002. Leg 183 synthesis: Kerguelen Plateau-Broken Ridge—a large igneous province. In Frey, F. A., Coffin, M. F., Wallace, P. J., and Quilty, P. G., eds., *Proc. ODP, Sci. Results.* College Station, TX: Ocean Drilling Program, vol. 183, pp. 1–48.

Fries, W. 1984. Cortical projections to the superior colliculus in the macaque monkey: a retrograde study using horseradish peroxidase. *J. Comp. Neurol.* 230: 55–76.

Frisby, J. P. 1980. *Seeing: Illusion, Brain and Mind.* New York: Oxford University Press.

Ganel, T., and Goodale, M. A. 2003. Visual control of action but not perception requires analytical processing of object shape. *Nature* 426: 664–667.

Garber, P. A., Blomquist, G. E., and Anzenberger, G. 2005. Kinematic analysis of trunk-to-trunk leaping in *Callimico goeldii*. *Int. J. Primatol.* 26: 223–240.

Garey, L. J., Dreher, B., and Robinson, S. R. 1991. The organization of the visual thalamus. In Dreher, B., and Robinson, S. R., eds., *Neuroanatomy of the Visual Pathways and Their Development*. Boca Raton, FL: CRC Press, pp. 176–234.

Gauld, I. D., and Wahl, D. B. 2002. The Eucerotinae: a Gondwanan origin for a cosmopolitan group of Ichneumonidae? *J. Nat. Hist.* 36: 2229–2248.

Gebo, D. L. 2002. Adapiformes: phylogeny and adaptation. In Hartwig, W. C., ed., *The Primate Fossil Record*. Cambridge: Cambridge University Press, pp. 21–43.

Gebo, D. L. 2004. A shrew-sized origin for primates. *Yrbk. Phys. Anthropol.* 47: 40–62.

Gifford, R., and Tristem, M. 2003. The evolution, distribution and diversity of endogenous retroviruses. *Virus Genes* 26: 291–315.

Gilad, Y., Wiebe, V., Przeworski, M., Lancet, D., and Pääbo, S. 2004. Loss of olfactory receptor genes coincides with the acquisition of full trichromatic vision in primates. *PloS Biol.* 2: 0120–0125.

Glaw, F., and Vences, M. 1994. *A Fieldguide to the Amphibians and Reptiles of Madagascar*, 2nd ed. Cologne, Germany: Verlags GbR.

Glendenning, K. K., Hall, J. A., Diamond, I. T., and Hall, W. C. 1975. The pulvinar nucleus of *Galago senegalensis*. *J. Comp. Neurol.* 161: 419–458.

Gloyd, H. K., and Conant, R. 1990. Snakes of the *Agkistrodon* complex: a monographic review. *Contr. Herpetol. No. 6. Soc. Study. Amph. Rept.* Miami, Ohio: Oxford.

Goldberg, A., Wildman, D. E., Schmidt, T. R., Hüttemann, M., Goodman, M., Weiss, M. L., and Grossman, L. I. 2003. Adaptive evolution of cytochrome *c* oxidase subunit VIII in anthropoid primates. *Proc. Nat. Acad. Sci.* 100: 5873–5878.

Goldin-Meadow, S. 2007. Pointing sets the stage for learning language—and creating language. *Child Develop.* 78: 741–745.

Goodale, M. A., and Milner, A. D. 1992. Separate visual pathways for perception and action. *Trends Neurosci.* 15: 20–25.

Goodale, M. A., and Westwood, D. A. 2004. An evolving view of duplex vision: separate but interacting cortical pathways for perception and action. *Curr. Opin. Neurobiol.* 14: 203–211.

Goodman, M. 1982. Positive selection causes purifying selection. *Nature* 295: 630.

Goossens, L., Schruers, K., Peeters, R., Griez, E., and Sunaert, S. 2007. Visual presentation of phobic stimuli: amygdala activation via an extrageniculostriate pathway? *Psychiat. Res.: Neuroimaging* 155: 113–120.

Gower, D. J., Kupfer, A., Oommen, O. V., Himstedt, W., Nussbaum, R. A., Loader, S. P., Presswell, B., Muller, H., Krishna, S. B., Boistel, R., and Wilkinson, M. 2002. A molecular phylogeny of ichthyophiid caecilians (Amphibia: Gymnophiona: Ichthyophiidae): out of India or out of South East Asia? *Proc. R. Soc. Lond. B* 269: 1563–1569.

Graham, A. 2003. Geohistory models and Cenozoic paleoenvironments of the Caribbean region. *System. Bot.* 28: 378–386.

Grant, L. 2001. *On a Kenya Ranch*. Perthshire, UK: Pioneer Associates.

Gravlund, P. 2001. Radiation within the advanced snakes (Caenophidia) with special emphasis on African opisthoglyph colubrids, based on mitochondrial sequence data. *Biol. J. Linn. Soc.* 72: 99–114.

Gray, D., Gutierrez, C., and Cusick, C. G. 1999. Neurochemical organization of inferior pulvinar complex in squirrel monkeys and macaques revealed by acetylcholinesterase histochemistry, calbindin and Cat-301 immunostaining, and *Wisteria floribunda* agglutinin binding. *J. Comp. Neurol.* 409: 452–468.

Greene, H. W. 1983. Dietary correlates of the origin and radiation of snakes. *Amer. Zool.* 23: 431–441.

Greene, H. W. 1997. *Snakes: The Evolution of Mystery in Nature*. Berkeley: University of California Press.

Greene, H. W., and Burghardt, G. M. 1978. Behavior and phylogeny: constriction in ancient and modern snakes. *Science* 200: 74–77.

Greene, H. W., and Cundall, D. 2000. Limbless tetrapods and snakes with legs. *Science* 287: 1939–1941.

Grieve, K. L., Acuña, C., and Cudeiro, J. 2000. The primate pulvinar nuclei: vision and action. *Trends Neurol. Sci.* 23: 35–39.

Griffiths, C. S. 1999. Phylogeny of the Falconidae inferred from molecular and morphological data. *Auk* 116: 116–130.

Griffiths, C. S., Barrowclough, G. F., Groth, J. G., and Mertz, L. A. 2004. Phylogeny of the Falconidae (Aves): a comparison of the efficacy of morphological, mitochondrial, and nuclear data. *Mol. Phyl. Evol.* 32: 101–109.

Griffiths, C. S., Barrowclough, G. F., Groth, J. G., and Mertz, L. A. 2007. Phylogeny, diversity, and classification of the Accipitridae based on DNA sequences of the RAG-1 exon. *J. Avian Biol.* 38: 587–602.

Grossman, L. I., Schmidt, T. R., Wildman, D. E., and Goodman, M. 2001. Molecular evolution of aerobic energy metabolism in primates. *Mol. Phylogen. Evol.* 18: 26–36.

Gursky, S. 2002. Predation on a wild spectral tarsier *(Tarsius spectrum)* by a snake. *Folia Primatol.* 73: 60–62.

Gursky, S. 2005. Predator mobbing in spectral tarsiers. *Int. J. Primatol.* 26: 207–236.

Gursky, S. 2006. Function of snake mobbing in spectral tarsiers. *Amer. J. Phys. Anthropol.* 129: 601–608.

Gutierrez, C., Cola, M. G., Seltzer, B., and Cusick, C. 2000. Neurochemical and connectional organization of the dorsal pulvinar complex in monkeys. *J. Comp. Neurol.* 419: 61–86.

Guyot, L. L., Diaz, F. G., O'Regan, M. H., Song, D., and Phillis, J. W. 2000. Topical glucose and accumulation of excitotoxic and other amino acids in ischemic cerebral cortex. *Horm. Metab. Res.* 32: 6–9.

Haberny, K. A., Paule, M. G., Scallet, A. C., Sistare, F. D., Lester, D. S., Hanig, J. P., and Slikker Jr., W. 2002. Ontogeny of the N-methyl-D-aspartate

(NMDA) receptor system and susceptibility to neurotoxicity. *Toxicol. Sci.* 68: 9–17.

Hackett, T. B., Wingfield, W. E., Mazzafaro, E. M., and Benedetti, J. S. 2002. Clinical findings associated with prairie rattlesnake bites in dogs: 100 cases (1989–1998). *J. Amer. Vet. Med. Assoc.* 220: 1675–1680.

Hammer, M., and Arlettaz, R. 1998. A case of snake predation upon bats in northern Morocco: some implications for designing bat grilles. *J. Zool., Lond.* 245: 211–212.

Hamrick, M. W. 2001. Primate origins: evolutionary change in digital ray patterning and segmentation. *J. Hum. Evol.* 49: 339–351.

Hankerson, S. J., and Caine, N. G. 2004. Pre-retirement predator encounters alter the morning behavior of captive marmosets *(Callithrix geoffroyi). Amer. J. Primatol.* 63: 75–85.

Hardy, D. L. 1994. *Bothrops asper* (Viperidae) snakebite and field researchers in middle America. *Biotropica* 26: 198–207.

Haring, E., Kruckenhauser, L., Gamauf, A., Riesing, M. J., and Pinsker, W. 2001. The complete sequence of the mitochondrial genome of *Buteo buteo* (Aves, Accipitridae) indicates an early split in the phylogeny of raptors. *Mol. Biol. Evol.* 18: 1892–1904.

Hart, D., and Sussman, R. W. 2005. *Man the Hunted: Primates, Predators, and Human Evolution.* New York: Westview Press.

Harting, J. K., Huerta, M. F., Frankfurther, A. J., Strominger, N. L., and Royce, G. J. 1980. Ascending pathways from the monkey superior colliculus: an autoradiographic analysis. *J. Comp. Neurol.* 192: 853–882.

Harting, J. K., Huerta, M. F., Hashikawa, T., and van Leishout, D. P. 1991. Projection of the mammalian superior colliculus upon the dorsal lateral geniculate nucleus: organization of tectogeniculate pathways in nineteen species. *J. Comp. Neurol.* 304: 275–306.

Harvey, P. H., Martin, R. D., and Clutton-Brock, T. H. 1987. Life histories in comparative perspective. In Smuts, B. B., Cheney, D. L., Seyfarth, R. M., Wrangham, R. W., and Struhsaker, T. T., eds., *Primate Societies.* Chicago: University of Chicago Press, pp. 181–196.

Hay, W. W., DeConto, R. M., Wold, C. N., Wilson, K. M., Voigt, S., Schulz, M., Wold, A. R., Dullo, W.-C., Ronov, A. B., Balukhovsky, A. N., and Söding, E. 1999. Alternative global Cretaceous paleogeography. In Berrera, E., and Johnson, C., eds., *Evolution of Cretaceous Ocean/Climate Systems. Geol. Soc. Amer. Spec. Paper* 332: 1–47.

Hayes, S. L., and Snowdon, C. T. 1990. Predator recognition in cotton-top tamarins *(Saguinus oedipus). Amer. J. Primatol.* 20: 283–291.

Hedges, S. B. 2006. Paleogeography of the Antilles and origin of West Indian terrestrial vertebrates. *Ann. Missouri Bot. Gard.* 93: 231–244.

Hedges, S. B., Parker, P. H., Sibley, C. G., and Kumar, S. 1996. Continental breakup and the ordinal diversification of birds and mammals. *Nature* 381: 226–229.

Heesy, C. P. 2005. Function of the mammalian postorbital bar. *J. Morphol.* 264: 363–380.

Heffner, R. S., Koay, G., and Heffner, H. E. 1999. Sound localization in an Old-World fruit bat *(Rousettus aegyptiacus)*: acuity, use of binaural cues, and relationship to vision. *J. Comp. Psychol.* 113: 297–306.

Heimel, J. A., Van Hooser, S. D., and Nelson, S. B. 2005. Laminar organization of response properties in primary visual cortex of the gray squirrel *(Sciurus carolinensis)*. *J. Neurophysiol.* 94: 3538–3554.

Heise, P. J., Maxson, L. R., Dowling, H. G., and Hedges, S. B. 1995. Higher-level snake phylogeny inferred from mitochondrial DNA sequences of 12S rRNA and 16S rRNA genes. *Mol. Biol. Evol.* 12: 259–265.

Hendry, S. H. C., and Reid, C. 2000. The koniocellular pathway in primate vision. *Ann. Rev. Neurosci.* 23: 127–153.

Hendry, S. H. C., and Yoshioka, T. 1994. A neurochemically distinct third channel in the macaque dorsal lateral geniculate nucleus. *Science* 264: 575–577.

Henneberry, R. C. 1989. The role of neuronal energy in the neurotoxicity of excitatory amino acids. *Neurobiol. Aging* 10: 611–613.

Henry, G. H., and Vidyasagar, T. R. 1991. Evolution of mammalian visual pathways. In Cronly-Dillon, J. R., and Gregory, R. L., eds., *Vision and Visual Dysfunction, Vol. 2: Evolution of the Eye and Visual System*. Boca Raton, FL: CRC Press, pp. 442–465.

Henschel, P. 2005. Leopard food habits in the Lopé National Park, Gabon, Central Africa. *Afr. J. Ecol.* 43: 21–28.

Hertz, L., Dringen, R., Schousboe, A., and Robinson, S. R. 1999. Astrocytes: glutamate producers for neurons. *J. Neurosci. Res.* 57: 417–428.

Hess, D. T., and Edwards, M. A. 1987. Anatomical demonstration of ocular segregation in the retinogeniculocortical pathway of the New World capuchin monkey *(Cebus apella)*. *J. Comp. Neurol.* 264: 409–420.

Heymann, E. W. 1987. A field observation of predation on a moustached tamarin *(Saguinus mystax)* by an anaconda. *Int. J. Primatol.* 8: 193–195.

Hooker, J. J., Russell, D. E., and Phélizon, A. 1999. A new family of Plesiadapiformes (Mammalia) from the Old World lower Paleogene. *Palaeontology* 42: 377–407.

Hopkins, W. D., Russell, J., Freeman, H., Buehler, N., Reynolds, E., and Schapiro, S. J. 2005. The distribution and development of handedness for manual gestures in captive chimpanzees *(Pan troglodytes)*. *Psychol. Sci.* 16: 487–493.

Hoppe-Dominik, B. 1984. Etude du spectre des proies de la panthère, *Panthera pardus,* dans le Parc National de Täi en Côte d'Ivoire. *Mammalia* 48: 477–487.

Horton, J. C. 1984. Cytochrome oxidase patches: a new cytoarchitectonic feature of monkey visual cortex. *Phil. Trans. R. Soc. Lond. B* 304: 199–253.

Horton, J. C., and Hedley-White, E. T. 1984. Mapping of cytochrome oxidase patches and ocular dominance columns in human visual cortex. *Phil. Trans. R. Soc. Lond. B* 304: 255–272.

Horton, J. C., and Hocking, D. R. 1996. Anatomical demonstration of ocular dominance columns in striate cortex of the squirrel monkey. *J. Neurosci.* 16: 5510–5522.

Hu, Y., Meng, J., Wang, Y., and Li, C. 2005. Large Mesozoic mammals fed on young dinosaurs. *Nature* 433: 149–152.

Hubel, D. H., and Livingstone, M. S. 1987. Segregation of form, color, and stereopsis in primate area 18. *J. Neurosci.* 7: 3378–3415.

Huchon, D., Madsen, O., Sibbald, M. J. J. B., Ament, K., Stanhope, M. J., Catzeflis, F., de Jong, W. W., and Douzery, E. J. P. 2002. Rodent phylogeny and a timescale for the evolution of Glires: evidence from an extensive taxon sampling using three nuclear genes. *Mol. Biol. Evol.* 19: 1053–1065.

Huerta, M. F., and Harting, J. K. 1983. Sublamination within the superficial gray layer of the squirrel monkey: an analysis of the tectopulvinar projection using anterograde and retrograde transport methods. *Brain Res.* 261: 119–126.

Huerta, M. F., and Harting, J. K. 1984. The mammalian superior colliculus: studies of its morphology and connections. In Vanegas, H., ed., *Comparative Neurology of the Optic Tectum.* New York: Plenum Press, pp. 687–773.

Hugall, A. F., Foster, R., and Lee, M. S. Y. 2007. Calibration choice, rate smoothing, and the pattern of tetrapod diversification according to the long nuclear gene RAG-1. *Syst. Biol.* 56: 542–563.

Hunt, D. M., Dulai, K. S., Cowing, J. A., Julliot, C., Mollon, J. D., Bowmaker, J. K., Li, W.-H., and Hewett-Emmett, D. 1998. Molecular evolution of trichromacy in primates. *Vis. Res.* 38: 3299–3306.

Iacoboni, M., and Mazziotta, J. C. 2007. Mirror neuron system: basic findings and clinical applications. *Ann. Neurol.* 62: 213–218.

Iacoboni, M., Molnar-Szakacs, I., Gallese, V., Buccino, G., Mazziotta, J. C., and Rizzolatti, G. 2005. Grasping the intentions of others with one's own mirror neuron system. *PLoS Biol.* 3: 529–535.

Ibbotson, M. R., and Mark, R. F. 2003. Orientation and spatiotemporal tuning of cells in the primary visual cortex of an Australian marsupial, the wallaby *Macropus eugenii. J. Comp. Physiol.* A 189: 115–123.

Iglesias, D. J., Tadeo, F. R., Legaz, F., Primo-Millo, E., and Talon, M. 2001. In vivo sucrose stimulation of colour change in citrus fruit epicarps: interactions between nutritional and hormonal signals. *Physiol. Plant.* 112: 244–250.

Ignashchenkova, A., Dicke, P. W., Haarmeier, T., and Thier, P. 2004. Neuron-specific contribution of the superior colliculus to overt and covert shifts of attention. *Nature Neurosci.* 7: 56–64.

Irvin, G. E., Norton, T. T., Sesma, M. A., and Casagrande, V. A. 1986. W-like response properties of interlaminar zone cells in the lateral geniculate nucleus of a primate *(Galago crassicaudatus). Brain Res.* 362: 254–270.

Isa, T. 2002. Intrinsic processing in the mammalian superior colliculus. *Curr. Opin. Neurobiol.* 12: 668–677.

Isbell, L. A. 1990. Sudden short-term increase in mortality of vervet monkeys *(Cercopithecus aethiops)* due to leopard predation in Amboseli National Park, Kenya. *Amer. J. Primatol.* 21: 41–52.

Isbell, L. A. 1994a. The vervets' year of doom. *Nat. Hist.* 8: 48–55.

Isbell, L. A. 1994b. Predation on primates: ecological patterns and evolutionary consequences. *Evol. Anthropol.* 3: 61–71.

Isbell, L. A. 2004. Is there no place like home? Ecological bases of dispersal in primates and their consequences for the formation of kin groups. In Chapais, B., and Berman, C., eds., *Kinship and Behavior in Primates.* New York: Oxford University Press, pp. 71–108.

Isbell, L. A. 2006a. Snakes as agents of evolutionary change in primate brains. *J. Hum. Evol.* 51: 1–35.

Isbell, L. A. 2006b. Did primates originate in Indo-Madagascar? A new possibility based on molecules, plate tectonics, and paleobiogeography. *Int. J. Primatol.* 27 (S1): 114.

Isbell, L. A. 2006c. Snakes on the brain. *New York Times,* September 3, sec. 4, p. 10.

Isbell, L. A., and Enstam, K. L. 2002. Predator (in)sensitive foraging in sympatric female vervets *(Cercopithecus aethiops)* and patas monkeys *(Erythrocebus patas):* a test of ecological models of group dispersion. In Miller, L. E., ed., *Eat or Be Eaten: Predator Sensitive Foraging among Primates.* New York: Cambridge University Press, pp. 154–168.

Isbell, L. A., Pruetz, J. D., Lewis, M., and Young, T. P. 1998. Locomotor activity differences between sympatric patas monkeys *(Erythrocebus patas)* and vervet monkeys *(Cercopithecus aethiops):* implications for the evolution of long hindlimb length in *Homo. Amer. J. Phys. Anthropol.* 105: 199–207.

Isbell, L. A., and Van Vuren, D. 1996. Differential costs of locational and social dispersal and their consequences for female group-living primates. *Behaviour* 133: 1–36.

Iwai, E., and Yukie, M. 1987. Amygdalofugal and amygdalopetal connections with modality-specific visual cortical areas in macaques (*Macaca fuscata, M. mulatta,* and *M. fascicularis*). *J. Comp. Neurol.* 261: 362–387.

Iwaniuk, A. N., and Whishaw, I. Q. 2000. On the origin of skilled forelimb movements. *Trends Neurosci.* 23: 372–376.

Jack, K. M. 2007. The cebines: toward an explanation of variable social structure. In Campbell, C. J., Fuentes, A., MacKinnon, K. C., Panger, M., and Bearder, S. K., eds., *Primates in Perspective.* New York: Oxford University Press, pp. 107–123.

Jackson, S. R., Newport, R., Mort, D., and Husain, M. 2005. Where the eye looks, the hand follows: limb-dependent magnet misreaching in optic ataxia. *Curr. Biol.* 15: 42–46.

Jacobs, G. H. 1993. The distribution and nature of colour vision among the mammals. *Biol. Rev.* 68: 413–471.

Jacobs, G. H. 1995. Variations in primate color vision: mechanisms and utility. *Evol. Anthropol.* 3: 196–205.

Jacobs, G. H., and Deegan II, J. F. 1999. Uniformity of colour vision in Old World monkeys. *Proc. R. Soc. Lond. B* 266: 2023–2028.

Jacobs, G. H., and Deegan II, J. F. 2001. Photopigments and colour vision in New World monkeys from the family Atelidae. *Proc. R. Soc. Lond. B* 268: 695–702.

Jacobs, G. H., Deegan II, J. F., Tan, Y., and Li, W.-S. 2002. Opsin gene and pho-topigment polymorphism in a prosimian primate. *Vis. Res.* 42: 11–18.

Jacobs, G. H., Neitz, M., Deegan II, J. F., and Neitz, J. 1996. Trichromatic colour vision in New World monkeys. *Nature* 382: 156–158.

Jaeger, J.-J., Courtillot, V., and Tapponnier, P. 1989. Paleontological view of the ages of the Deccan Traps, the Cretaceous/Tertiary boundary, and the India-Asia collision. *Geology* 17: 316–319.

Jansa, S. A., Goodman, S. M., and Tucker, P. K. 1999. Molecular phylogeny and biogeography of the native rodents of Madagascar (Muridae: nesomyinae): a test of the single-origin hypothesis. *Cladistics* 15: 253–270.

Johnson, W. E., and Coffin, J. M. 1999. Constructing primate phylogenies from ancient retrovirus sequences. *Proc. Natl. Acad. Sci.* 96: 10254–10260.

Jones, E. G. 1985. *The Thalamus.* New York: Plenum Press.

Jones, E. G., and Burton, H. 1976. A projection from the medial pulvinar to the amygdala in primates. *Brain Res.* 104: 142–147.

Jones, F. W. 1916. *Arboreal Man.* London: Arnold.

Jones, G., and Teeling, E. C. 2006. The evolution of echolocation in bats. *Trends Ecol. Evol.* 21: 149–156.

Jörg, J., Jock, S., Boucsein, W., and Schäfer, F. 2004. Zur autonomen dysregula-tion beim freezing-phänomen von morbus-Parkinson-patienten: ein ambula-torisches monitoring und videorecording. *Aktuelle Neurol.* 31: 338–346.

Julesz, B. 1971. *Foundations of Cyclopean Perception.* Chicago: University of Chicago Press.

Kaas, J. H. 2004. Early visual areas: V1, V2, V3, DM, DL, and MT. In Kaas, J. H., and Collins, C. E., eds., *The Primate Visual System.* New York: CRC Press, pp. 139–159.

Kaas, J. H., and Huerta, M. F. 1988. The subcortical visual system of primates. In Steklis, H. D., and Erwin, J., eds., *Comparative Primate Biology,* vol. 4. New York: Alan R. Liss, pp. 327–391.

Kaas, J. H., Huerta, M. F., Weber, J. T., and Harting, J. K. 1978. Patterns of ret-inal terminations and laminar organization of the lateral geniculate nucleus of primates. *J. Comp. Neurol.* 182: 517–554.

Kadenbach, B., Hüttemann, M., Arnold, S., Lee, I., and Bender, E. 2000. Mito-chondrial energy metabolism is regulated via nuclear-controlled subunits of cytochrome *c* oxidase. *Free Radical Biol. Med.* 29: 211–221.

Kadoya, S., Wolin, L. R., and Massopust Jr., L. C. 1971. Photically evoked unit activity in the tectum opticum of the squirrel monkey. *J. Comp. Neurol.* 142: 495–508.

Kainz, P. M., Neitz, J., and Neitz, M. 1998. Recent evolution of trichromacy in a New World monkey. *Vis. Res.* 38: 3315–3320.

Kalin, N. H., Shelton, S. E., and Davidson, R. J. 2004. The role of the central nu-cleus of the amygdala in mediating fear and anxiety in the primate. *J. Neu-rosci.* 24: 5506–5515.

Kalin, N. H., Shelton, S. E., Davidson, R. J., and Kelley, A. E. 2001. The primate amygdala mediates acute fear but not the behavioral and physiological com-ponents of anxious temperament. *J. Neurosci.* 21: 2067–2074.

Kalko, E. K. V., Herre, E. A., and Handley Jr., C. O. 1996. Relation of fig fruit characteristics to fruit-eating bats in the New and Old World tropics. *J. Biogeog.* 23: 565–576.

Kaplan, E. 2004. The M, P, and K pathways of the primate visual system. In Chalupa, L. M., and Werner, J. S., eds., *The Visual Neurosciences.* Cambridge, MA: MIT Press, pp. 481–493.

Kaplan, G., and Rogers, L. J. 2006. Head-cocking as a form of exploration in the common marmoset and its development. *Develop. Psychobiol.* 48: 551–560.

Kappeler, P. M. 2000. Lemur origins: rafting by groups of hibernators? *Folia Primatol.* 71: 422–425.

Kardong, K. V. 2002. Colubrid snakes and Duvernoy's "venom" glands. *J. Toxicol.: Toxin Rev.* 21: 1–19.

Kastner, S., De Weerd, P., and Ungerleider, L. G. 2000. Texture segregation in the human visual cortex: a functional MRI study. *J. Neurophysiol.* 83: 2453–2457.

Kastner, S., O'Connor, D. H., Fukui, M. M., Fehd, H. M., Herwig, U., and Pinsk, M. A. 2004. Functional imaging of the human lateral geniculate nucleus and pulvinar. *J. Neurophysiol.* 91: 438–448.

Kawashima, R., Sugiura, M., Kato, T., Nakamura, A., Hatano, K., Ito, K., Fukuda, H., Kojima, S., and Nakamura, K. 1999. The human amygdala plays an important role in gaze monitoring: a PET study. *Brain* 122: 779–783.

Kay, R. F., and Kirk, E. C. 2000. Osteological evidence for the evolution of activity pattern and visual acuity in primates. *Amer. J. Phys. Anthropol.* 113: 235–262.

Kay, R. F., Williams, B. A., Ross, C. F., Takai, M., and Shigehara, N. 2004. Anthropoid origins: a phylogenetic analysis. In Ross, C., and Kay, R. F., eds., *Anthropoid Origins: New Visions.* New York: Kluwer Academic/Plenum Press, pp. 91–135.

Keogh, J. S. 1998. Molecular phylogeny of elapid snakes and a consideration of their biogeographic history. *Bio. J. Linn. Soc.* 63: 177–203.

Khajuria, C. K., and Prasad, G. V. R. 1998. Taphonomy of a Late Cretaceous mammal-bearing microvertebrate assemblage from the Deccan inter-trappean beds of Naskal, peninsular India. *Palaeogeog., Palaeoclimat., Palaeoecol.* 137: 153–172.

Kiltie, R. A. 2000. Scaling of visual acuity with body size in mammals and birds. *Funct. Ecol.* 14: 226–234.

King, S. M., and Cowey, A. 1992. Defensive responses to looming visual stimuli in monkeys with lateral striate cortex ablation. *Neuropsychology* 30: 1017–1024.

Kingdon, J. 1997. *The Kingdon Field Guide to African Mammals.* New York: Academic Press.

Kirk, E. C. 2006a. Effects of activity pattern on eye size and orbital aperture size in primates. *J. Hum. Evol.* 51: 159–170.

Kirk, E. C. 2006b. Eye morphology in cathemeral lemurids and other mammals. *Folia Primatol.* 77: 27–49.

Kirk, E. C., Cartmill, M., and Kay, R. F. 2003. Comment on "grasping primate origins." *Science* 300: 741b.

Kirk, E. C., and Kay, R. F. 2004. The evolution of high visual acuity in the Anthropoidea. In Ross, C., and Kay, R. F., eds. *Anthropoid Origins: New Visions.* New York: Kluwer Academic/Plenum Press, pp. 539–602.

Kluge, A. G. 1991. Boine snake phylogeny and research cycles. *Misc. Publ. Mus. Zool., U. Michigan* 178: 1–58.

Knight, A., and Mindell, D. P. 1994. On the phylogenetic relationship of Colubrinae, Elapinae, and Viperidae and the evolution of front-fanged venom systems in snakes. *Copeia* 1: 1–9.

Knott, C. D., and Kahlenberg, S. M. 2007. Orangutans in perspective: forced copulations and female mating resistance. In Campbell, C. J., Fuentes, A., MacKinnon, K. C., Panger, M., and Bearder, S. K., eds., *Primates in Perspective.* New York: Oxford University Press, pp. 290–305.

Kobatake, E., and Tanaka, K. 1994. Neuronal selectivities to complex object features in the ventral visual pathway of the macaque cerebral cortex. *J. Neurophysiol.* 71: 856–867.

Kovács, G., Vogels, R., and Orban, G. A. 1995a. Cortical correlate of pattern backward masking. *Proc. Natl. Acad. Sci.* 92: 5587–5591.

Kovács, G., Vogels, R., and Orban, G. A. 1995b. Selectivity of macaque inferior temporal neurons for partially occluded shapes. *J. Neurosci.* 15: 1984–1997.

Krane, S., Itagaki, Y., Nakanishi, K., and Weldon, P. J. 2003. "Venom" of the slow loris: sequence similarity of prosimian skin gland protein and FEl d 1 cat allergen. *Naturwissenschaften* 90: 60–62.

Krause, D. W., O'Connor, P. M., Rogers, K. C., Sampson, S. D., Buckley, G. A., and Rogers, R. R. 2006. Late Cretaceous terrestrial vertebrates from Madagascar: implications for Latin American biogeography. *Ann Missouri Bot. Gard.* 93: 178–208.

Krause, D. W., Prasad, G. V. R., von Koenigswald, W., Sahni, A., and Grine, F. E. 1997. Cosmopolitanism among Gondwanan Late Cretaceous mammals. *Nature* 390: 504–507.

Krauzlis, R. J., Basso, M. A., and Wurtz, R. H. 2000. Discharge properties of neurons in the rostral superior colliculus of the monkey during smooth-pursuit eye movements. *J. Neurophysiol.* 84: 876–891.

Kress, W. J. 1993. Coevolution of plants and animals: pollination of flowers by primates in Madagascar. *Curr. Sci.* 65: 253–257.

Kress, W. J., Schatz, G. E., Andrianifahanana, M., and Simons Morland, H. 1994. Pollination of *Ravenala madagascariensis* (Strelitziaceae) by lemurs in Madagascar: evidence of an archaic coevolutionary system? *Amer. J. Bot.* 81: 542–551.

Krubitzer, L. A. 1998. What can monotremes tell us about brain evolution? *Phil. Trans. R. Soc. Lond. B* 353: 1127–1146.

Krubitzer, L. A., and Kaas, J. H. 1990. Cortical connections of MT in four species of primates: areal, modular, and retinotopic patterns. *Vis. Neurosci.* 5: 165–204.

Kudo, H., and Dunbar, R. I. M. 2001. Neocortex size and social network size in primates. *Anim. Behav.* 62: 711–722.

Kumar, S., and Hedges, S. B. 1998. A molecular timescale for vertebrate evolution. *Nature* 392: 917–919.

Künzle, H. 1996. Diencephalic connections of the superior colliculus in the hedgehog tenrec. *Exp. Brain Res.* 111: 356–370.

LaBerge, D., and Buchsbaum, M. S. 1990. Positron emission tomographic measurements of pulvinar activity during an attention task. *J. Neurosci.* 10: 613–619.

Lachica, E. A., Beck, P. D., and Casagrande, V. A. 1993. Intrinsic connections of layer III of striate cortex in squirrel monkey and bush baby: correlations with patterns of cytochrome oxidase. *J. Comp. Neurol.* 329: 163–187.

Lachica, E. A., and Casagrande, V. A. 1992. Direct W-like geniculate projections to the cytochrome oxidase (CO) blobs in primate visual cortex: axon morphology. *J. Comp. Neurol.* 319: 141–158.

Lachica, E. A., and Casagrande, V. A. 1993. The morphology of collicular and retinal axons ending on small relay (W-like) cells of the primate lateral geniculate nucleus. *Vis. Neurosci.* 10: 403–418.

Lane, R. H., Allman, J. M., and Kaas, J. H. 1971. Representation of the visual field in the superior colliculus of the grey squirrel *(Sciurus carolinensis)* and the tree shrew *(Tupaia glis)*. *Brain Res.* 26: 277–292.

Langton, S. R. H., and Bruce, V. 2000. You *must* see the point: automatic processing of cues to the direction of social attention. *J. Exp. Psychol.* 26: 747–757.

Langton, S. R. H., Watt, R. J., and Bruce, V. 2000. Do the eyes have it? Cues to the direction of social attention. *Trends Cogn. Sci.* 4: 50–59.

Leavens, D. A., Hopkins, W. D., and Bard, K. A. 2005. Understanding the point of chimpanzee pointing. *Curr. Dir. Psychol. Sci.* 14: 185–189.

LeDoux, J. 2000. The amygdala and emotion: a view through fear. In Aggleton, J. P., ed., *The Amygdala: A Functional Analysis*. New York: Oxford University Press, pp. 281–310.

LeDoux, J. 2002. *Synaptic Self: How Our Brains Become Who We Are*. New York: Penguin Books.

Lee, M. S. Y., Bell Jr., G. L., and Caldwell, M. W. 1999. The origin of snake feeding. *Nature* 400: 655–659.

Lee, M. S. Y., and Scanlon, J. D. 2002. Snake phylogeny based on osteology, soft anatomy and ecology. *Biol. Rev.* 77: 333–402.

Leeming, D. 1994. *A Dictionary of Creation Myths*. New York: Oxford University Press.

Le Gros Clark, W. E. 1941. The lateral geniculate body in the platyrrhine monkeys. *J. Anat.* 76: 131–140.

Le Gros Clark, W. E. 1959. *The Antecedents of Man*. Edinburgh: Edinburgh University Press.

Lenk, P., Kalyabina, S., Wink, M., and Joger, U. 2001. Evolutionary relationships among the true vipers (Reptilia: Viperidae) inferred from mitochondrial DNA sequences. *Mol. Phylogen. Evol.* 19: 94–104.

Levine, S., Atha, K., and Wiener, S. G. 1993. Early experience effects on the development of fear in the squirrel monkey. *Behav. Neural Biol.* 60: 225–233.

Levitt, J. B., Yoshioka, T., and Lund, J. S. 1995. Connections between the pulvinar complex and cytochrome oxidase-defined compartments in visual area V2 of macaque monkey. *Exp. Brain Res.* 104: 419–430.

Li, W.-H., and Grauer, D. 1997. *Fundamentals of Molecular Evolution.* Sunderland, MA: Sinauer Associates.

Liddell, B. S., Brown, K. J., Kemp, A. H., Barton, M. J., Das, P., Peduto, A., Gordon, E., and Williams, M. 2005. A direct brainstem-amygdala-cortical "alarm" system for subliminal signals of fear. *NeuroImage* 24: 235–243.

Lieb, K., Brucker, S., Bach, M., Els, T., Lücking, C. H., and Greenlee, M. W. 1999. Impairment in preattentive visual processing in patients with Parkinson's disease. *Brain* 122: 303–313.

Liszkowski, U., Carpenter, M., Henning, A., Striano, T., and Tomasello, M. 2004. Twelve-month-olds point to share attention and interest. *Develop. Sci.* 7: 297–307.

Livingstone, M. S., and Hubel, D. H. 1982. Thalamic inputs to cytochrome oxidase-rich regions in monkey visual cortex. *Proc. Natl. Acad. Sci.* 79: 6098–6101.

Lucas, D. R., and Newhouse, J. P. 1957. The toxic effects of sodium L-glutamate on the inner layers of the retina. *Arch. Ophthalmol.* 58: 193–201.

Lucas, P. W., Darvell, B. W., Lee, P. K. D., Yen, T. D. B., and Choong, M. F. 1998. Colour cues for leaf food selection by long-tailed macaques *(Macaca fascicularis)* with a new suggestion for the evolution of trichromatic colour vision. *Folia Primatol.* 69: 139–152.

Lünenburger, L., Kutz, D. F., and Hoffmann, K.-P. 2000. Influence of arm movements on saccades in humans. *Eur. J. Neurosci.* 12: 4107–4116.

Maas, M. C., Krause, D. W., and Strait, S. G. 1988. The decline and extinction of Plesiadapiformes (Mammalia: ?Primates) in North America: displacement or replacement? *Paleobiology* 14: 410–431.

Macartney, M. J., Gregory, P. T., and Larsen, K. W. 1988. A tabular survey of data on movements and home ranges of snakes. *J. Herpetol.* 22: 61–73.

Macey, J. R., Schulte, J. A., Larson, A., Ananjeva, N. B., Wang, Y., Pethiyagoda, R., Rastegar-Pouyani, N., and Papenfuss, T. J. 2000. Evaluating trans-Tethys migration: an example using Acrodont lizard phylogenetics. *Syst. Biol.* 49: 233–256.

Machado, C. A., Jousselin, E., Kjellberg, F., Compton, S. G., and Herre, E. A. 2001. Phylogenetic relationships, historical biogeography and character evolution of fig-pollinating wasps. *Proc. R. Soc. Lond. B* 268: 685–694.

Madsen, O., Scalley, M., Douady, C. J., Kao, D. J., DeBry, R. W., Adkins, R., Amrine, H. M., Stanhope, M. J., de Jong, W. W., and Springer, M. S. 2001. Parallel adaptive radiations in two major clades of placental mammals. *Nature* 409: 610–614.

Marivaux, L., Bocat, L., Chaimanee, Y., Jaeger, J.-J., Marandat, B., Srisuk, P., Tafforeau, P., Yamee, C., and Welcomme, J.-L. 2006. Cynocephalid dermopterans from the Paleogene of South Asia (Thailand, Myanmar and

Pakistan): systematic, evolutionary and palaeobiogeographic implications. *Zool. Scripta* 35: 395–420.

Marrocco, R. T. 1972. Sustained and transient cells in monkey lateral geniculate nucleus: conduction velocities and response properties. *J. Neurophysiol.* 39: 340–353.

Martin, G. R., and Katzir, G. 1999. Visual fields in short-toed eagles, *Circaetus gallicus* (Accipitridae), and the function of binocularity in birds. *Brain Behav. Evol.* 53: 55–66.

Martin, L. D. 1989. Fossil history of the terrestrial carnivora. In Gittleman, J. L., ed., *Carnivore Behavior, Ecology, and Evolution.* Ithaca, NY: Cornell University Press, pp. 536–568.

Martin, P. R., White, J. J. R., Goodchild, A. K., Wilder, H. D., and Sefton, A. E. 1997. Evidence that blue-on cells are part of the third geniculocortical pathway in primates. *Eur. J. Neurosci.* 9: 1536–1541.

Martin, R. D. 1986. Vertebrate phylogeny: are fruit bats primates? *Nature* 320: 482–483.

Martin, R. D. 1990. *Primate Origins and Evolution: A Phylogenetic Reconstruction.* Princeton, NJ: Princeton University Press.

Martin, R. D. 1993. Primate origins: plugging the gaps. *Nature* 363: 223–234.

Martin, R. D. 1996. Scaling of the mammalian brain: the maternal energy hypothesis. *News Physiol. Sci.* 11: 149–156.

Martin, R. D. 2000. Origins, diversity and relationships of lemurs. *Int. J. Primatol.* 21: 1021–1049.

Martin, R. D. 2003. Combing the primate record. *Nature* 422: 388–391.

Martin, R. D. 2006. New light on primate evolution. *Berichte und Abhandlungen der Berlin-Brandenburgische Akademie der Wissenschaften* 11: 379–405.

Martin, R. D., Soligo, C., and Tavaré, S. 2007. Primate origins: implications of a Cretaceous ancestry. *Folia Primatol.* 78: 277–296.

Masters, J. C., Anthony, N. M., de Wit, M. J., and Mitchell, A. 2005. Reconstructing the evolutionary history of the Lorisidae using morphological, molecular, and geological data. *Amer. J. Phys. Anthropol.* 127: 465–480.

Maunsell, J. H. R. 1987. Physiological evidence for two visual subsystems. In Vaina, L. M., ed., *Matters of Intelligence: Conceptual Structures in Cognitive Neuroscience.* Boston: Reidel, pp. 59–87.

Maunsell, J. H. R., and Newsome, W. T. 1987. Visual processing in monkey extrastriate cortex. *Ann. Rev. Neurosci.* 10: 363–401.

McCall, R. A. 1997. Implications of recent geological investigations of the Mozambique Channel for the mammalian colonization of Madagascar. *Proc. R. Soc. Lond. B* 264: 663–665.

McDonald, A. J. 1998. Cortical pathways to the mammalian amygdala. *Prog. Neurobiol.* 55: 257–332.

McDowell, S., and Harris, J. 1997. Irrelevant peripheral visual stimuli impair manual reaction times in Parkinson's disease. *Vis. Res.* 37: 3549–3558.

McGaugh, J. L., Ferry, B., Vazdarjanova, A., and Roozendaal, B. 2000. Amygdala: role in modulation of memory storage. In Aggleton, J. P., ed., *The*

Amygdala: A Functional Analysis. New York: Oxford University Press, pp. 391–423.

McKenna, M. C., and Bell, S. K. 1997. *Classification of Mammals above the Species Level.* New York: Columbia University Press.

McNamee, G. 2000. *The Serpent's Tale: Snakes in Folklore and Literature.* Athens: University of Georgia Press.

Meldrum, B., and Garthwaite, J. 1990. Excitatory amino acid neurotoxicity and neurodegenerative disease. *Trends Pharmacol. Sci.* 11: 379–387.

Melmoth, D. R., and Grant, S. 2006. Advantages of binocular vision for the control of reaching and grasping. *Exp. Brain Res.* 171: 371–388.

Meng, J. 2004. Phylogeny and divergence of basal Glires. *Bull. Amer. Mus. Nat. Hist.* 285: 93–109.

Menzel, C. R. 1980. Head-cocking and visual perception in primates. *Anim. Behav.* 28: 151–159.

Menzel, C. R., and Menzel Jr., E. W. 1980. Head-cocking and visual exploration in marmosets *(Saguinus fuscicollis). Behaviour* 75: 219–233.

Merigan, W. H., Nealey, T. A., and Maunsell, J. H. R. 1993. Visual effects of lesions in cortical area V2 in macaques. *J. Neurosci.* 13: 3180–3191.

Metz, M., Piliponsky, A. M., Chen, C.-C., Lammel, V., Åbrink, M., Pejler, G., Tsai, M., and Galli, S. J. 2006. Mast cells can enhance resistance to snake and honeybee venoms. *Science* 313: 526–530.

Michaux, J., and Catzeflis, F. 2000. The bushlike radiation of muroid rodents is exemplified by the molecular phylogeny of the LCAT nuclear gene. *Mol. Phylogen. Evol.* 17: 280–293.

Miller, E. R., Gunnell, G. F., and Martin, R. D. 2005. Deep time and the search for anthropoid origins. *Yrbk. Phys. Anthropol.* 48: 60–95.

Miller, L. E. 2002. The role of group size in predator sensitive foraging decisions in wedge-capped capuchin monkeys *(Cebus olivaceus).* In Miller, L. E., ed., *Eat or Be Eaten: Predator Sensitive Foraging among Primates.* Cambridge: Cambridge University Press, pp. 95–106.

Miller, L. E., and Treves, A. 2007. Predation on primates: past studies, current challenges, and directions for the future. In Campbell, C. J., Fuentes, A., MacKinnon, K. C., Panger, M., and Bearder, S. K., eds., *Primates in Perspective.* New York: Oxford University Press, pp. 525–543.

Milton, K. 1980. *The Foraging Strategy of Howler Monkeys: A Study in Primate Economics.* New York: Columbia University Press.

Milton, K. 1988. Foraging behaviour and the evolution of primate intelligence. In Byrne, R., and Whiten, A., eds., *Machiavellian Intelligence.* Oxford: Oxford University Press, pp. 285–305.

Mirtschin, P. J., Masci, P., Paton, D. C., and Kuchel, T. 1998. Snake bites recorded by veterinary practices in Australia. *Aust. Vet. J.* 76: 195–198.

Mitchell, I. J., Dean, P., and Redgrave, P. 1988. The projection from superior colliculus to cuneiform area in the rat. *Exp. Brain Res.* 72: 626–639.

Mitchell, P. C., and Pocock, R. I. 1907. On the feeding of reptiles in captivity with observations on the fear of snakes by other vertebrates. *Proc. Zool. Soc. Lond.* 2: 785–794.

Mohr, B. A. R., Wähnert, V., and Lazarus, D. 2002. Mid-Cretaceous paleobotany and palynology of the central Kerguelen Plateau, Southern Indian Ocean (ODP leg 183, site 1138). In Frey, F. A., Coffin, M. F., Wallace, P. J., and Quilty, P. G., eds., *Proc. ODP, Sci. Results.* College Station, TX: Ocean Drilling Program, vol. 183, pp. 1–39.

Møller, A. P. 1997. Parasitism and the evolution of host life-history. In Clayton, D. H., and Moore, J., eds., *Host-Parasite Evolution: General Principles and Avian Models.* New York: Oxford University Press, pp. 105–127.

Mollon, J. D. 1989. "Tho' she kneel'd in that place where they grew . . .": the uses and origins of primate colour vision. *J. Exp. Biol.* 146: 21–38.

Montje, M. L., Phillips, R., and Sapolsky, R. 2001. Calbindin overexpression buffers hippocampus cultures from the energetic impairments caused by glutamate. *Brain Res.* 911: 37–42.

Morales, M., Mundy, P., Delgado, C. E. F., Yale, M., Messinger, D., Neal, R., and Schwartz, H. 2000. Responding to joint attention across the 6- through 24-month age period and early language acquisition. *J. Appl. Develop. Psychol.* 21: 283–298.

Morand, S., Thut, G., de Peralta, R. G., Clarke, S., Khateb, A., Landis, T., and Michel, C. M. 2000. Electrophysiological evidence for fast visual processing through the human koniocellular pathway when stimuli move. *Cerebral Cortex* 10: 817–825.

Morgan, J. J., Adam, A., and Mollon, J. D. 1992. Dichromats detect colour-camouflaged objects that are not detected by trichromats. *Proc. R. Soc. Lond. B* 248: 291–295.

Morley, R. J. 2000. *Origin and Evolution of Tropical Rain Forests.* New York: John Wiley and Sons.

Morley, R. J., and Dick, C. W. 2003. Missing fossils, molecular clocks, and the origin of the Melastomataceae. *Amer. J. Bot.* 90: 1638–1644.

Morris, J. S., deBonis, M., and Dolan, R. J. 2002. Human amygdala responses to fearful eyes. *NeuroImage* 17: 214–222.

Morris, J. S., Friston, K. J., Büchel, C., Frith, C. D., Young, A. W., Calder, A. J., and Dolan, R. J. 1998. A neuromódulatory role for the human amygdala in processing emotional facial expressions. *Brain* 121: 47–57.

Morris, J. S., Friston, K. J., and Dolan, R. J. 1997. Neural responses to salient visual stimuli. *Proc. R. Soc. Lond. B* 264: 769–775.

Morris, J. S., Öhman, A., and Dolan, R. J. 1999. A subcortical pathway to the right amygdala mediating "unseen" fear. *Proc. Natl. Acad. Sci.* 96: 1680–1685.

Morrison, J. H., and Foote, S. L. 1986. Noradrenergic and serotonergic innervation of cortical, thalamic, and tectal visual structures in Old and New World monkeys. *J. Comp. Neurol.* 243: 117–138.

Müller, B., Goodman, S. M., and Peichl, L. 2007. Cone photoreceptor diversity in the retinas of fruit bats (Megachiroptera). *Brain Behav. Evol.* 70: 90–104.

Müller, K.-H., Ahl, C., and Hartmann, G. 1997. Geophagy in masked titi monkeys *(Callicebus personatus melanochir)* in Brazil. *Primates* 38: 69–71.

Mundy, P., Sigman, M., Ungerer, J., and Sherman, T. 1986. Defining the social deficits of autism: the contribution of non-verbal communication measures. *J. Child Psycho. Psychiat.* 27: 657–669.

Munoz, D. P., and Wurtz, R. H. 1995. Saccade-related activity in monkey superior colliculus. I. Characteristics of burst and buildup cells. *J. Neurophysiol.* 73: 2313–2333.

Murphy, K. M., Jones, D. G., and Van Sluyters, R. C. 1995. Cytochrome-oxidase blobs in cat primary visual cortex. *J. Neurosci.* 15: 4196–4208.

Murphy, W. J., Eizirik, E., Johnson, W. E., Zhang, Y. P., Ryder, O. A., and O'Brien, S. J. 2001a. Molecular phylogenetics and the origins of placental mammals. *Nature* 409: 614–618.

Murphy, W. J., Eizirik, E., O'Brien, S. J., Madsen, O., Scally, M., Douady, C. J., Teeling, E., Ryder, O. A., Stanhope, M. J., de Jong, W. W., and Springer, M. S. 2001b. Resolution of the early placental mammal radiation using Bayesian phylogenetics. *Science* 294: 2348–2351.

Nagy, A., Kruse, W., Rottmann, S., Dannenberg, S., and Hoffmann, K.-P. 2006. Somatosensory-motor neuronal activity in the superior colliculus of the primate. *Neuron* 52: 525–534.

Napier, J. R., and Walker, A. C. 1967. Vertical clinging and leaping—a newly recognized category of locomotor behavior of primates. *Folia Primatol.* 6: 204–219.

Nascimento-Silva, S., Gattass, R., Fiorani Jr., M., and Sousa, A. P. B. 2003. Three streams of visual information processing in V2 of *Cebus* monkey. *J. Comp. Neurol.* 466: 104–118.

Neggers, S. F. W., and Bekkering, H. 2002. Coordinated control of eye and hand movements in dynamic reaching. *Hum. Move. Sci.* 21: 349–376.

Nei, M., and Glazko, G. V. 2002. Estimation of divergence times for a few mammalian and several primate species. *J. Hered.* 93: 157–164.

Nekaris, K. A. I. 2005. Foraging behaviour of the slender loris *(Loris lydekkerianus lydekkerianus)*: implications for theories of primate origins. *J. Hum. Evol.* 49: 289–300.

Neophytou, S. I., Aspley, S., Butler, S., Beckett, S., and Marsden, C. A. 2001. Effects of lesioning noradrenergic neurones in the locus coeruleus on conditioned and unconditioned aversive behaviour in the rat. *Prog. Neuro-Psychopharmacol. Biol. Psychiat.* 25: 1307–1321.

Nie, F., and Wong-Riley, M. T. T. 1995. Double labeling of GABA and cytochrome oxidase in the macaque visual cortex: quantitative EM analysis. *J. Comp. Neurol.* 356: 115–131.

Nie, F., and Wong-Riley, M. T. T. 1996. Differential glutamatergic innervation in cytochrome oxidase-rich and -poor regions of the macaques striate cortex: quantitative EM analysis of neurons and neuropil. *J. Comp. Neurol.* 369: 571–590.

Niemitz, C., ed. 1984. *The Biology of Tarsiers.* New York: Gustave Fischer Verlag.

Nieuwboer, A., Dom, R., De Weerdt, W., Desloovere, K., Fieuws, S., and Broens-Kaucsik, E. 2001. Abnormalities of the spatiotemporal characteristics of gait at the onset of freezing in Parkinson's disease. *Movement Disorders* 16: 1066–1075.

Noble, V. E., Kowalski, E. M., and Ravosa, M. J. 2000. Orbit orientation and the function of the mammalian postorbital bar. *J. Zool. Lond.* 250: 405–418.

Noonan, B. P., and Chippindale, P. T. 2006. Dispersal and vicariance: the complex evolutionary history of boid snakes. *Mol. Phylogen. Evol.* 40: 347–358.

Northmore, O. P. M., Levine, E. S., and Schneider, G. E. 1988. Behavior evoked by electrical stimulation of the hamster superior colliculus. *Exp. Brain Res.* 73: 595–605.

Norton, T. T., and Casagrande, V. A. 1982. Laminar organization of receptive-field properties in lateral geniculate nucleus of bush baby *(Galago crassicaudatus)*. *J. Neurophysiol.* 47: 715–741.

Novelli, A., Reilly, J. A., Lysko, P. G., and Henneberry, R. C. 1988. Glutamate becomes neurotoxic via the N-methyl-D-aspartate receptor when intracellular energy levels are reduced. *Brain Res.* 451: 205–212.

Ogren, M. P., and Hendrickson, A. E. 1977. The distribution of pulvinar terminals in visual areas 17 and 18 of the monkey. *Brain Res.* 137: 343–350.

Öhman, A. 2007. Has evolution primed humans to "beware the beast"? *Proc. Nat. Acad. Sci.* 104: 16396–16397.

Öhman, A., Flykt, A., and Esteves, F. 2001. Emotion drives attention: detecting the snake in the grass. *J. Exp. Psychol: General* 130: 466–478.

Öhman, A., and Mineka, S. 2003. The malicious serpent: snakes as a prototypical stimulus for an evolved module of fear. *Curr. Dir. Psychol. Sci.* 12: 5–9.

Öhman, A., and Soares, J. J. F. 1993. On the automatic nature of phobic fear: conditioned electrodermal responses to masked fear-relevant stimuli. *J. Abnorm. Psychol.* 102: 121–132.

Öhman, A., and Soares, J. J. F. 1994. "Unconscious anxiety": phobic responses to masked stimuli. *J. Abnorm. Psychol.* 103: 231–240.

Okusa, T., Kakigi, R., and Osaka, N. 2000. Cortical activity related to cue-invariant shape perception in humans. *Neuroscience* 98: 615–624.

Olney, J. W. 1969. Brain lesions, obesity and other disturbances in mice treated with monosodium glutamate. *Science* 164: 719–721.

Olney, J. W. 1990. Excitotoxic amino acids and neuropsychiatric disorders. *Ann. Rev. Pharmacol. Toxicol.* 30: 47–71.

Olson, L. E., Sargis, E. J., and Martin, R. D. 2005. Intraordinal phylogenetics of treeshrews (Mammalia: Scandentia) based on evidence from the mitochondrial 12S rRNA gene. *Mol. Phylogen. Evol.* 35: 656–673.

Orrego, F., and Villanueva, S. 1993. The chemical nature of the main central excitatory neurotransmitter: a critical appraisal based on release studies and synaptic vesicle localization. *Neuroscience* 56: 539–555.

Ouchi, Y., Yoshikawa, E., Okada, H., Futatsubashi, M., Sekine, Y., Iyo, M., and Sakamoto, M. 1999. Alterations in binding site density of dopamine transporter in the striatum, orbitofrontal cortex, and amygdala in early Parkinson's disease: compartment analysis for β-CFT binding with positron emission tomography. *Ann. Neurol.* 45: 601–610.

Ovid. 1916. *Metamorphoses,* vol. 1, trans. F. J. Miller. Cambridge, MA: Harvard University Press.

Padian, K., and Chiappe, L. M. 1998. The origin and early evolution of birds. *Biol. Rev.* 73: 1–42.

Parker, A. 2003. *In the Blink of an Eye.* Cambridge, MA: Perseus Publishing.

Parkinson, C. L. 1999. Molecular systematics and biogeographical history of pitvipers as determined by mitochondrial ribosomal DNA sequences. *Copeia* 3: 576–586.

Pérez, J. C., and Sánchez, E. E. 1999. Natural protease inhibitors to hemorrhagins in snake venoms and their potential use in medicine. *Toxicon* 37: 703–728.

Perret, M., Aujard, F., Séguy, M., and Schilling, A. 2003. Olfactory bulbectomy modifies photic entrainment and circadian rhythms of body temperature and locomotor activity in a nocturnal primate. *J. Biol. Rhythms* 18: 392–401.

Perry, V. H., and Cowey, A. 1984. Retinal ganglion cells that project to the superior colliculus and pretectum in the macaque monkey. *Neuroscience* 12: 1125–1137.

Pessiglione, M., Guehl, D., Agid, Y., Hirsch, E. C., Féger, J., and Tremblay, L. 2003. Impairment of context-adapted movement selection in a primate model of presymptomatic Parkinson's disease. *Brain* 126: 1392–1408.

Peterhans, E. 1997. Functional organization of area V2 in the awake monkey. In Rockland, K. S., Kaas, J. H., and Peters, A., eds., *Cerebral Cortex, Vol. 12: Extrastriate Cortex in Primates.* New York: Plenum Press, pp. 295–333.

Peterhans, E., and von der Heydt, R. 1993. Functional organization of area V2 in the alert macaque. *Europ. J. Neurosci.* 5: 509–524.

Pettigrew, J. D., Jamison, B. G. M., Robson, S. K., Hall, L. S., McAnally, K. I., and Cooper, H. M. 1989. Phylogenetic relations between microbats, megabats and primates (Mammalia: Chiroptera and Primates). *Phil. Trans. R. Soc. Lond. B* 325: 489–559.

Pitkänen, A. 2000. Connectivity of the rat amygdaloid complex. In Aggleton, J. P., ed., *The Amygdala: A Functional Analysis.* New York: Oxford University Press, pp. 31–115.

Polly, P. D., Wesley-Hunt, G. D., Heinrich, R. E., Davis, G., and Houde, P. 2006. Earliest known carnivoran auditory bulla and support for a recent origin of crown-group Carnivora (Eutheria, Mammalia). *Paleontology* 49: 1019–1027.

Polyak, S. 1957. *The Vertebrate Visual System.* Chicago: University of Chicago Press.

Popke, E. J., Allen, R. R., Pearson, E. C., Hammond, T. G., and Paule, M. G. 2001. Differential effects of two NMDA receptor antagonists on cognitive-behavioral performance in young nonhuman primates II. *Neurotox. Teratol.* 23: 333–347.

Poran, N. S., Coss, R. G., and Benjamini, E. 1987. Resistance of California ground squirrels *(Spermophilus beecheyi)* to the venom of the Northern Pacific rattlesnake *(Crotalus viridis oreganus):* a study of adaptive variation. *Toxicon* 25: 767–777.

Poux, C., Madsen, O., Marquard, E., Vieites, D. R., de Jong, W. W., and Vences, M. 2005. Asynchronous colonization of Madagascar by the four endemic clades of primates, tenrecs, carnivores, and rodents as inferred from nuclear genes. *Syst. Biol.* 54: 719–730.

Povinelli, D. J., and Davis, D. R. 1994. Differences between chimpanzees *(Pan troglodytes)* and humans in the resting state of the index finger: implications for pointing. *J. Comp. Psychol.* 108: 134–139.

Prasad, G. V. R., Jaeger, J. J., Sahni, A., Gheerbrant, E., and Khajuria, C. K. 1994. Eutherian mammals from the Upper Cretaceous (Maastrichtian) intertrappean beds of Naskal, Andrah Pradesh, India. *J. Vert. Paleontol.* 14: 260–277.

Prescott, J. J., and Buchanan-Smith, H. M. 2002. Predation sensitive foraging in captive tamarins. In Miller, L. E., ed., *Eat or Be Eaten: Predator Sensitive Foraging among Primates.* Cambridge: Cambridge University Press, pp. 44–57.

Preuss, T. M. 2007. Evolutionary specializations of primate brain systems. In Ravoso, M. J., and Dagosto, M., eds., *Primate Origins and Adaptations.* New York: Kluwer Academic/Plenum Press, pp. 625–675.

Preuss, T. M., Beck, P. D., and Kaas, J. H. 1993. Areal, modular, and connectional organization of visual cortex in a prosimian primate, the slow loris *(Nycticebus coucang). Brain Behav. Evol.* 42: 321–335.

Preuss, T. M., and Kaas, J. H. 1996. Cytochrome oxidase "blobs" and other characteristics of primary visual cortex in a lemuroid primate, *Cheirogaleus medius. Brain Behav. Evol.* 47: 103–112.

Previc, F. H. 1990. Functional specialization in the lower and upper visual fields in humans: its ecological origins and neurophysiological implications. *Behav. Brain Sci.* 13: 519–575.

Qiu, F., Sugihara, T., and von der Heydt, R. 2007. Figure-ground mechanisms provide structure for selective attention. *Nature Neurosci.* 10: 1492–1499.

Quammen, D. 2006. *The Reluctant Mr. Darwin.* New York: W. W. Norton.

Quiroga, R. Q., Reddy, L., Kreiman, G., Koch, C., and Fried, I. 2005. Invariant visual representation by single neurons in the human brain. *Nature* 435: 1102–1107.

Rafal, R., Smith, J., Krantz, J., Cohen, A., and Brennan, C. 1990. Extrageniculate vision in hemianopic humans: saccade inhibition by signals in the blind field. *Science* 250: 118–121.

Rage, J.-C. 1996. Les Madtsoiidae (Reptilia, Serpentes) du Crétacé supérieur d'Europe: témoins gondwaniens d'une dispersion transtéthysienne. *C. R. Acad. Sci. Paris* 322: 603–608.

Rage, J.-C. 2003. Relationships of the Malagasy fauna during the Late Cretaceous: northern or southern routes? *Acta Palaeontol. Polonica* 48: 661–662.

Rage, J.-C., Bajpai, S., Thewissen, J. G. M., and Tiwari, B. N. 2003. Early Eocene snakes from Kutch, Western India, with a review of the Palaeophiidae. *Geodiversitas* 25: 695–716.

Rage, J.-C., Cappetta, H., Hartenberger, J.-L., Jaeger, J.-J., Sudre, J., Vianey-Liaud, M., Kumar, K., Prasad, G. V. R., and Sahni, A. 1995. Collision age. *Nature* 375: 286.

Rage, J.-C., and Escuillié, F. 2000. Un nouveau serpent bipède du Cénomanien (Crétacé). Implications phylétiques. *Earth Planet. Sci.* 330: 513–520.

Rage, J.-C., Prasad, G. V. R., and Bajpai, S. 2004. Additional snakes from the uppermost Cretaceous (Maastrichtian) of India. *Cretaceous Res.* 25: 425–434.

Rage, J.-C., and Werner, C. 1999. Mid-Cretaceous (Cenomanian) snakes from Wadi Abu Hashim, Sudan: the earliest snake assemblage. *Palaeont. Afr.* 35: 85–110.

Ramakrishnan, U., Coss, R. G., Schank, J., Dharawat, A., and Kim, S. 2005. Snake species discrimination by wild bonnet macaques *(Macaca radiata)*. *Ethology* 111: 337–356.

Randall, J. A., Hatch, S. M., and Hekkela, E. R. 1995. Interspecific variation in antipredator behavior in sympatric species of kangaroo rat. *Behav. Ecol. Sociobiol.* 36: 243–250.

Rasmussen, D. T. 2002a. The origin of primates. In Hartwig, W. C., ed., *The Primate Fossil Record.* Cambridge: Cambridge University Press, pp. 5–10.

Rasmussen, D. T. 2002b. Early catarrhines of the African Eocene and Oligocene. In Hartwig, W. C., ed., *The Primate Fossil Record.* Cambridge: Cambridge University Press, pp. 203–220.

Ravosa, M. J., and Savakova, D. G. 2004. Euprimate origins: the eyes have it. *J. Hum. Evol.* 46: 357–364.

Rawlings, L. H., Rabosky, D. L., Donnellan, S. C., and Hutchinson, M. N. 2008. Python phylogenetics: inference from morphology and mitochondrial DNA. *Biol. J. Linn. Soc.* 93: 603–619.

Regan, B. C., Julliot, C., Simmen, B., Viénot, F., Charles-Dominique, P., and Mollon, J. D. 1998. Frugivory and colour vision in *Alouatta seniculus,* a trichromatic platyrrhine monkey. *Vis. Res.* 38: 3321–3327.

Regan, B. C., Julliot, C., Simmen, B., Viénot, F., Charles-Dominique, P., and Mollon, J. D. 2001. Fruits, foliage and the evolution of primate colour vision. *Phil. Trans. R. Soc. Lond. B* 356: 229–283.

Reid, R. C., Alonso, J.-M., and Hendry, S. H. C. 1997. S-cone input is relayed to visual cortex from two koniocellular layers of macaque LGN. *Soc. Neurosci. Abstr.* 23: 13.

Reyes, A., Gissi, C., Catzeflis, F., Nevo, E., Pesole, G., and Saccone, C. 2004. Congruent mammalian trees from mitochondrial and nuclear genes using Bayesian methods. *Mol. Biol. Evol.* 21: 397–403.

Rhoades, R. W., Mooney, R. D., and Fish, S. E. 1991. Retinotopic and visuotopic representations in the mammalian superior colliculus. In Dreher, B., and Robinson, S. R., eds., *Neuroanatomy of the Visual Pathways and Their Development.* Boca Raton, FL: CRC Press, pp. 150–175.

Riba-Hernandez, P., Stoner, K. E., and Lucas, P. W. 2003. The sugar composition of fruits in the diet of spider monkeys *(Ateles geoffroyi)* in tropical humid forest in Costa Rica. *J. Trop. Biol.* 19: 709–716.

Riba-Hernandez, P., Stoner, K. E., and Lucas, P. W. 2005. Sugar concentration of fruits and their detection via color in the Central American spider monkey *(Ateles geoffroyi). Amer. J. Primatol.* 67: 411–423.

Richard, A. F. 1985. *Primates in Nature*. New York: W. H. Freeman.

Riesing, J. J., Kruckenhauser, L., Gamauf, A., and Haring, E. 2003. Molecular phylogeny of the genus *Buteo* (Aves: Accipitridae) based on mitochondrial marker sequences. *Mol. Phylogen. Evol.* 27: 328–342.

Rizzolatti, G., and Matelli, M. 2003. Two different streams in the dorsal visual system: anatomy and function. *Exp. Brain Res.* 153: 146–157.

Roberts, A. D., Shelton, S. E., Barnhart, T. E., Oakes, T. R., Davidson, R. J., and Kalin, N. H. 2002. Effects of snake exposure on brain glucose metabolism in the rhesus monkey. Paper presented at the Society for Neuroscience 32nd annual meeting, Orlando, Florida.

Robinson, D. L. 1993. Functional contributions of the primate pulvinar. In Hicks, T. P., Molotchinikoff, S., and Ono, T., eds., *Progress in Brain Research*, vol. 95. Cambridge: Elsevier, pp. 371–380.

Robinson, D. L., and Petersen, S. E. 1992. The pulvinar and visual salience. *Trends Neurosci.* 15: 127–132.

Rodda, G. H., and Savidge, J. A. 2007. Biology and impacts of Pacific Island invasive species. 2. *Boiga irregularis,* the brown tree snake (Reptilia: Colubridae). *Pacific Sci.* 61: 307–324.

Rodman, H. R., Sorenson, K. M., Shim, A. J., and Hexter, D. P. 2001. Calbindin immunoreactivity in the geniculo-extrastriate system of the macaque: implications for heterogeneity in the koniocellular pathway and recovery from cortical damage. *J. Comp. Neurol.* 431: 168–181.

Rodriguez-Robles, J. A., Bell, C. J., and Greene, H. W. 1999. Gape size and evolution of diet in snakes: feeding ecology of erycine boas. *J. Zool.* 248: 49–58.

Roe, A. W. 2004. Modular complexity of area V2 in the macaque monkey. In Kaas, J. H., and Collins, C. E., eds., *The Primate Visual System*. New York: CRC Press, pp. 109–138.

Roe, A. W., and Ts'o, D. Y. 1995. Visual topography in primate V2: multiple representation across functional stripes. *J. Neurosci.* 15: 3689–3715.

Roe, A. W., and Ts'o, D. Y. 1997. The functional architecture of area V2 in the macaque monkey. In Rockland, K. S., Kaas, J. H., and Peters, A., eds., *Cerebral Cortex, Vol. 12: Extrastriate Cortex in Primates*. New York: Plenum Press, pp. 295–333.

Rogers, R. R., Hartman, J. H., and Krause, D. W. 2000. Stratigraphic analysis of Upper Cretaceous rocks in the Mahajanga Basin, Northwestern Madagascar: implications for ancient and modern faunas. *J. Geol.* 108: 275–301.

Rohen, J. W., and Castenholz, A. 1967. Über die zentralisation der retina bei primaten. *Folia Primatol.* 5: 92–147.

Romano, C., Price, M., Bai, H. Y., and Olney, J. W. 1993. Neuroprotectants in Honghua: glucose attenuates retinal ischemic damage. *Invest. Ophthal. Vis. Sci.* 34: 72–80.

Roos, C., Schmitz, J., and Zischler, H. 2004. Primate jumping genes elucidate strepsirrhine phylogeny. *Proc. Nat. Acad. Sci.* 101: 10650–10654.

Rosa, M. G. P., and Krubitzer, L. A. 1999. The evolution of visual cortex: where is V2? *Trends Neurosci.* 22: 242–248.

Rosa, M. G. P., and Schmid, L. M. 1994. Topography and extent of visual-field representation in the superior colliculus of the megachiropteran *Pteropus*. *Vis. Neurosci.* 11: 1037–1057.

Rose, K. D., DeLeon, V. B., Missiaen, P., Rana, R. S., Sahni, A., Singh, L., and Smith, T. 2008. Early Eocene lagomorph (Mammalia) from Western India and the early diversification of Lagomorpha. *Proc. R. Soc. Lond. B* 275: 1203–1208.

Rose, L. M., Perry, S., Panger, M. A., Jack, K., Manson, J. H., Gros-Louis, J., Mackinnon, K. C., and Vogel, E. 2003. Interspecific interactions between *Cebus capucinus* and other species: data from three Costa Rican sites. *Int. J. Primatol.* 24: 759–796.

Rosenberg, D. 1994. *World Mythology*, 3rd ed. Lincolnwood, IL: NTC Publishing.

Rosenbluth, D., and Allman, J. M. 2002. The effect of gaze angle and fixation distance on the responses of neurons in V1, V2, and V4. *Neuron* 33: 143–149.

Ross, C. F. 2000. Into the light: the origin of Anthropoidea. *Annu. Rev. Anthropol.* 29: 147–194.

Ross, C. F., and Kay, R. F. 2004. Anthropoid origins: retrospective and prospective. In Ross, C. F., and Kay, R. F., eds., *Anthropoid Origins: New Visions*. New York: Kluwer Academic/Plenum Press, pp. 701–737.

Rowling, J. K. 1997. *Harry Potter and the Sorcerer's Stone*. New York: Scholastic.

Rutschmann, F., Eriksonn, T., Schonenberger, J., and Conti, E. 2004. Did Crypteroniaceae really disperse out of India? Molecular dating evidence from rbcL, ndhF, and rp/16 intron sequences. *Int. J. Plant Sci.* 164: S69–S83.

Rydell, J., and Speakman, J. R. 1995. Evolution of nocturnality in bats: potential competitors and predaors during their early history. *Biol. J. Linn. Soc.* 54: 183–191.

Sahraie, A., Weiskrantz, L., Barbur, J. L., Simmons, A., and Williams, S. C. R. 1997. Pattern of neuronal activity associated with conscious and unconscious processing of visual signals. *Proc. Natl. Acad. Sci.* 94: 9406–9411.

Saito, A., Mikami, A., Kawamura, S., Ueno, Y., Hiramatsu, C., Widayati, K. A., Suryobroto, B., Teramoto, M., Mori, Y., Nagano, K., Fujita, K., Kuroshima, H., and Hasegawa, T. 2005. Advantage of dichromats over trichromats in discrimination of color-camouflaged stimuli in nonhuman primates. *Amer. J. Primatol.* 67: 425–436.

Sampson, S. D., Witmer, L. M., Forster, C. A., Krause, D. W., O'Connor P. M., Dodson, P., and Ravoavy, F. 1998. Predatory remains from Madagascar: implications for the Cretaceous biogeography of Gondwana. *Science* 280: 1048–1051.

Sanchez, M., Camara, M., Prohens, J., Ruiz, J. J., Torija, E., and Nuez, F. 2000. Variation in carbohydrate content during ripening in two clones of pepino. *J. Sci. Food Agric.* 80: 1985–1991.

Sargis, E. J. 2004. New views on tree shrews: the role of tupaiids in primate supraordinal relationships. *Evol. Anthropol.* 13: 56–66.

Scanlon, J. D. 2003. The basicranial morphology of madtsoiid snakes (Squamata, Ophidia) and the earliest Alethinophidia (Serpentes). *J. Vert. Paleontol.* 23: 971–976.

Scanlon, J. D. 2006. Skull of the large non-macrostomatan snake Yurlunggur from the Australian Oligo-Miocene. *Nature* 439: 839–842.

Scanlon, J. D., and Lee, M. S. Y. 2000. The Pleistocene serpent Wonambi and the early evolution of snakes. *Nature* 403: 416–420.

Scerif, G., Gomez, J.-C., and Byrne, R. W. 2004. What do Diana monkeys know about the focus of attention of a conspecific? *Anim. Behav.* 68: 1239–1247.

Schettino, L. F., Adamovich, S. V., Hening, W., Tunik, E., Sage, J., and Poizner, H. 2006. Hand preshaping in Parkinson's disease: effects of visual feedback and medication state. *Exp. Brain Res.* 168: 186–202.

Schettino, L. F., Rajaraman, V., Jack, D., Adamovich, S. V., Sage, J., and Poizner, H. 2003. Deficits in the evolution of hand preshaping in Parkinson's disease. *Neuropsychology* 42: 82–94.

Schneider, H., Canavez, F. C., Sampaio, I., Moreira, M. Â. M., Tagliaro, C. H., and Seuánez, H. N. 2001. Can molecular data place each neotropical monkey in its own branch? *Chromosoma* 109: 515–523.

Schneider, H., Schneider, M. P. C., Sampaio, I., Harada, M. L., Stanhope, M., Czelusniak, J., and Goodman, M. 1993. Molecular phylogeny of the New World monkeys (Platyrrhini, Primates). *Mol. Phylogen. Evol.* 2: 225–242.

Schrago, C. G., and Russo, C. A. M. 2003. Timing the origin of New World monkeys. *Mol. Biol. Evol.* 20: 1620–1625.

Schulz, B., Fendt, M., and Schnitzler, H.-U. 2002. Clonidine injections into the lateral nucleus of the amygdala block acquisition and expression of fear-potentiated startle. *Europ. J. Neurosci.* 15: 151–157.

Seiffert, E. R. 2006. Revised age estimates for the later Paleogene mammal faunas of Egypt and Oman. *Proc. Natl. Acad. Sci.* 103: 5000–5005.

Seiffert, E. R., Simons, E. L., and Attia, Y. 2003. Fossil evidence for an ancient divergence of lorises and galagos. *Nature* 422: 421–424.

Selemon, L. D., and Goldman-Rakic, P. S. 1988. Common cortical and subcortical targets of the dorsolateral prefrontal and posterior parietal cortices in the rhesus monkey: evidence for a distributed neural network subserving spatially guided behavior. *J. Neurosci.* 8: 4049–4068.

Seo, S. Y., Kim, E. Y., Kim, H., and Gwag, B. J. 1999. Neuroprotective effect of high glucose against NMDA, free radical, and oxygen-glucose deprivation through enhanced mitochondrial potentials. *J. Neurosci.* 19: 8849–8855.

Sewards, T. V., and Sewards, M. A. 2002. Innate visual object recognition in vertebrates: some proposed pathways and mechanisms. *Comp. Biochem. Physiol. Part A* 132: 861–891.

Seyfarth, R. M., Cheney, D. L., and Marler, P. 1980. Monkey responses to three different alarm calls: evidence for predator classification and semantic communication. *Science* 210: 801–803.

Shine, R., Harlow, P. S., Keogh, J. S., and Boeadi. 1998. The influence of sex and body size on food habits of a giant tropical snake, *Python reticulatus*. *Func. Ecol.* 12: 248–258.

Shipp, S., and Zeki, S. 1995. Segregation and convergence of specialized pathways in macaque monkey visual cortex. *J. Anat.* 187: 547–562.

Shostak, Y., Ding, Y., Mavity-Hudson, J., and Casagrande, V. A. 2002. Cortical synaptic arrangements of the third visual pathway in three primate species: *Macaca mulatta, Saimiri sciureus,* and *Aotus trivirgatus. J. Neurosci.* 22: 2885–2893.

Shulman, G. L., Ollinger, J. M., Linenweber, M., Petersen, S. E., and Corbetta, M. 2001. Multiple neural correlates of detection in the human brain. *Proc. Nat. Acad. Sci.* 98: 313–318.

Sibley, C. G., and Ahlquist, J. E. 1990. *Phylogeny and Classification of Birds: A Study in Molecular Evolution.* New Haven, CT: Yale University Press.

Siemers, B. M., Goerlitz, H. R., Robsomanitrandrasana, E., Piep, M., Ramanamanjato, J.-P., Rakotondravony, D., Ramilijaona, O., and Ganzhorn, J. U. 2007. Sensory basis of food detection in wild *Microcebus murinus. Int. J. Primatol.* 28: 291–304.

Sigé, B., Jaeger, J. J., Sundre, J., and Vianey-Liaud, M. 1990. *Altiatlasius koulchii* n. gen. et sp., primate omomyidé du Paléocène supérieur du Maroc, et les origines des euprimates. *Palaeontographica Abt. A.* 214: 31–56.

Silk, J. B., and Stanford, C. B. 1999. Infanticide article disputed. *Anthropol. News* 40: 27–29.

Simpson, G. G. 1940. Mammals and land bridges. *J. Wash. Acad. Sci.* 30: 137–163.

Sincich, L. C., and Horton, J. C. 2002. Divided by cytochrome oxidase: a map of the projections from V1 to V2 in macaques. *Science* 295: 1734–1737.

Sincich, L. C., and Horton, J. C. 2005. The circuitry of V1 and V2: integration of color, form, and motion. *Annu. Rev. Neurosci.* 28: 303–326.

Sincich, L. C., Park, K. F., Wohlgemuth, M. J., and Horton, J. C. 2004. Bypassing V1: a direct geniculate input to area MT. *Nature Neurosci.* 7: 1123–1128.

Slowinski, J. B., and Lawson, R. 2002. Snake phylogeny: evidence from nuclear and mitochondrial genes. *Mol. Phylogen. Evol.* 24: 194–202.

Smith, A. C., Buchanan-Smith, H. M., Surridge, A. K., Osorio, D., and Mundy, N. I. 2003. The effect of colour vision status on the detection and selection of fruits by tamarins (*Saguinus* spp.). *J. Exp. Biol.* 206: 3159–3165.

Smith, T., Rose, K. D., and Gingerich, P. D. 2006. Rapid Asia–Europe–North America geographic dispersal of earliest Eocene primate *Teilhardina* during the Paleocene-Eocene thermal maximum. *Proc. Nat. Acad. Sci.* 103: 11223–11227.

Smith, T., Van Itterbeeck, J., and Missiaen, P. 2004. Oldest plesiadapiform (Mammalia, Proprimates) from Asia and its palaeobiogeographical implications for faunal interchange with North America. *C. R. Palevol.* 3: 43–52.

Snyder, L. H., Calton, J. L., Dickinson, A. R., and Lawrence, B. M. 2002. Eye-hand coordination: saccades are faster when accompanied by a coordinated arm movement. *J. Neurophysiol.* 87: 2279–2286.

Soares, J. G. M., Diogo, A. C. M., Fiorani, M., Souza, A. P. B., and Gattass, R. 2001a. Changes in orientation and direction selectivity of cells in secondary visual area (V2) after GABA inactivation of the pulvinar in Cebus monkeys. *Soc. Neurosci. Abstr.* 27: 1633.

Soares, J. G. M., Gattass, R., Souza, A. P. B., Rosa, M. G. P., Fiorani Jr., M., and Brandão, B. L. 2001b. Connectional and neurochemical subdivisions of the pulvinar in *Cebus* monkeys. *Vis. Neurosci.* 18: 25–41.

Sparks, J. S. 2004. Molecular phylogeny and biogeography of the Malagasy and South Asian cichlids (Telostei: Perciformes: Cichlidae). *Mol. Phylogenet. Evol.* 30: 599–614.

Spawls, S., Howell, K., Drewes, R., and Ashe, J. 2002. *A Field Guide to the Reptiles of East Africa.* San Diego, CA: Academic Press.

Springer, M. S., Murphy, W. J., Eizirik, E., and O'Brien, S. J. 2003. Placental mammal diversification and the Cretaceous-Tertiary boundary. *Proc. Natl. Acad. Sci.* 100: 1056–1061.

Springer, M. S., Stanhope, M. J., Madsen, O., and de Jong, W. W. 2004. Molecules consolidate the placental mammal tree. *Trends Ecol. Evol.* 19: 430–438.

Stanford, C. B. 1989. Predation on capped langurs *(Presybitis pileata)* by cooperatively hunting jackals *(Canis aureus). Amer. J. Primatol.* 19: 53–56.

Stanhope, M. J., Waddell, V. G., Madsen, O., de Jong, W., Hedges, S. B., Cleven, G. C., Kao, D., and Springer, M. S. 1998. Molecular evidence for multiple origins of Insectivora and for a new order of endemic African insectivore mammals. *Proc. Natl. Acad. Sci.* 95: 9967–9972.

Stankiewicz, J., Thiart, C., Masters, J. C., and de Wit, M. J. 2006. Did lemurs have sweepstakes tickets? An exploration of Simpson's model for the colonization of Madagascar by mammals. *J. Biogeogr.* 33: 221–235.

Steenbeek, R., Piek, R. C., van Buul, M., and van Hooff, J. A. R. A. M. 1999. Vigilance in wild Thomas's langurs *(Presbytis thomasi):* the importance of infanticide risk. *Behav. Ecol. Sociobiol.* 45: 137–150.

Stepniewska, I. 2004. The pulvinar complex. In Kaas, J. H., and Collins, C. E., eds., *The Primate Visual System.* New York: CRC Press, pp. 53–80.

Stepniewska, I., Fang, P.-C., and Kaas, J. H. 2005. Microstimulation reveals specialized subregions for different complex movements in posterior parietal cortex of prosimians galagos. *Proc. Natl. Acad. Sci.* 102: 4878–4883.

Stepniewska, I., and Kaas, J. H. 1997. Architectonic subdivisions of the inferior pulvinar in New World and Old World monkeys. *Vis. Neurosci.* 14: 1043–1060.

Stepniewska, I., Qi, H.-X., and Kaas, J. H. 1999. Do superior colliculus projection zones in the inferior pulvinar project to MT in primates? *Europ. J. Neurosci.* 11: 469–480.

Stepniewska, I., Qi, H.-X., and Kaas, J. H. 2000. Projections of the superior colliculus to subdivisions of the inferior pulvinar in New World and Old World monkeys. *Vis. Neurosci.* 17: 529–549.

Stoerig, P., and Cowey, A. 1993. Blindsight: neurons and behaviour. In Hicks, T. P., Molotchnikoff, S., and Ono, T., eds., *Progress in Brain Research,* vol. 95. Cambridge: Elsevier, pp. 445–459.

Stoerig, P., Kleinschmidt, A., and Frahm, J. 1997. No visual responses in denervated V1: high-resolution functional magnetic resonance imaging of a blindsight patient. *NeuroReport* 9: 21–25.

Stone, J., and Johnston, E. 1981. The topography of primate retina: a study of the human, bushbaby, and New- and Old-World monkeys. *J. Comp. Neurol.* 196: 205–223.

Storey, M., Mahoney, J. J., Saunders, A. D., Duncan, R. A., Kelley, S. P., and Coffin, M. F. 1995. Timing of hot spot–related volcanism and the breakup of Madagascar and India. *Science* 267: 852–855.

Strenge, H. 1978. The functional significance of the pulvinar. *Fortschr. Neurol. Psychiat.* 46: 491–507.

Stuphorn, V., Bauswein, E., and Hoffmann, K.-P. 2000. Neurons in the primate superior colliculus coding for arm movements in gaze-related coordinates. *J. Neurophysiol.* 83: 1283–1299.

Sumner, P., and Mollon, J. D. 2000a. Catarrhine photopigments are optimized for detecting targets against a foliage background. *J. Exp. Biol.* 203: 1963–1986.

Sumner, P., and Mollon, J. D. 2000b. Chromaticity as a signal of ripeness in fruits taken by primates. *J. Exp. Biol.* 203: 1987–2000.

Surridge, A. K., Osorio, D., and Mundy, N. I. 2003. Evolution and selection of trichromatic vision in primates. *Trends Ecol. Evol.* 18: 198–205.

Sussman, R. W. 1991. Primate origins and the evolution of angiosperms. *Amer. J. Primatol.* 23: 209–223.

Tabuce, R., Mahboubi, M., Tafforeau, P., and Sudre, J. 2004. Discovery of a highly-specialized plesiadapiform primate in the early-middle Eocene of northwestern Africa. *J. Hum. Evol.* 47: 305–321.

Takai, M., Anaya, F., Shigehara, N., and Setoguchi, T. 2000. New fossil materials of the earliest New World monkey, *Branisella boliviana*, and the problem of platyrrhine origins. *Amer. J. Phys. Anthropol.* 111: 263–281.

Tan, Y., and Li, W.-H. 1999. Trichromatic vision in prosimians. *Nature* 402: 36.

Tan, Y., Yoder, A. D., Yamashita, N., and Li, W.-H. 2005. Evidence from opsin genes rejects nocturnality in ancestral primates. *Proc. Natl. Acad. Sci.* 102: 14712–14716.

Tanaka, K., Saito, H., Fukada, Y., and Moriya, M. 1991. Coding visual images of objects in the inferotemporal cortex of the macaque monkey. *J. Neurophysiol.* 66: 170–189.

Tavaré, S., Marshall, C. R., Will, O., Soligo, C., and Martin, R. D. 2002. Using the fossil record to estimate the age of the last common ancestor of extant primates. *Nature* 416: 726–729.

Taylor, H. 1999. What we are afraid of. The Harris Poll #49, August 18. www .harrisinteractive.com/harris_poll/index.asp?PID=281.

Teeling, E. C., Springer, M. S., Madsen, O., Bates, P., O'Brien, S. J., and Murphy, W. J. 2005. A molecular phylogeny for bats illuminates biogeography and the fossil record. *Science* 307: 580–584.

Tello, N. S., Huck, M., and Heymann, E. W. 2002. *Boa constrictor* attack and successful group defence in moustached tamarins, *Saguinus mystax*. *Folia Primatol.* 73: 146–148.

Terborgh, J. 1983. *Five New World Primates: A Study in Comparative Ecology.* Princeton, NJ: Princeton University Press.

Terborgh, J., and Janson, C. H. 1986. The socioecology of primate groups. *Ann. Rev. Ecol. Syst.* 17: 111–135.

Thewissen, J. G. M. 1990. Comment on "Paleontological view of the ages of the Deccan Traps, the Cretaceous/Tertiary boundary, and the India-Asia collision." *Geology* 18: 185.

Thiele, A., Vogelsang, M., and Hoffmann, K. P. 1991. Pattern of retinotectal projection in the megachiropteran bat *Rousettus aegyptiacus. J. Comp. Neurol.* 314: 671–683.

Thong, H. S. 1997. Out of mother Africa: how humans learned to think, speak and make culture. *Vietnam Rev.* 3: 520–560.

Tomasello, M. 2000. Culture and cognitive development. *Curr. Dir. Psychol. Sci.* 9: 37–40.

Tomasello, M., Carpenter, M., and Liszkowski, U. 2007. A new look at infant pointing. *Child Develop.* 78: 705–722.

Tootell, R. B. H., Silverman, M. S., De Valois, R. L., and Jacobs, G. H. 1983. Functional organization of the second cortical visual area in primates. *Science* 220: 737–739.

Trojanowski, J. Q., and Jacobson, S. 1974. Medial pulvinar afferents to frontal eye fields in rhesus monkey demonstrated by horseradish peroxidase. *Brain Res.* 80: 395–411.

Ts'o, D. Y., and Gilbert, C. D. 1988. The organization of chromatic and spatial interactions in the primate striate cortex. *J. Neurosci.* 8: 1712–1727.

Ts'o, D. Y., Roe, A. W., and Gilbert, C. D. 2001. A hierarchy of the functional organization for color, form and disparity in primate visual area V2. *Vis. Res.* 41: 1333–1349.

Tsukahara, T. 1993. Lions eat chimpanzees: the first evidence of predation by lions on wild chimpanzees. *Amer. J. Primatol.* 29: 1–11.

Ungerleider, L. G., and Christensen, C. A. 1979. Pulvinar lesions in monkeys produce abnormal scanning of a complex visual array. *Neurpsychologia* 17: 493–501.

van der Kuyl, A. C., Dekker, J. T., and Goudsmit, J. 2000. Primate genus *Miopithecus:* evidence for the existence of species and subspecies of dwarf guenons based on cellular and endogenous viral sequences. *Mol. Phylog. Evol.* 14: 403–413.

Vandoros, J. D., and Dumont, E. R. 2004. Use of the wings in manipulative and suspensory behaviors during feeding by frugivorous bats. *J. Exp. Zool., Part A: Comp. Exp. Biol.* 301A: 361–366.

Van Hooser, S. D., and Nelson, S. B. 2006. The squirrel as a rodent model of the human visual system. *Vis. Neurosci.* 23: 765–778.

van Schaik, C. P. 1983. Why are diurnal primates living in groups? *Behaviour* 87: 120–144.

van Schaik, C. P., and Kappeler, P. M. 1996. The social systems of gregarious lemurs: lack of convergence with anthropoids due to evolutionary disequilibrium? *Ethology* 102: 915–941.

van Schaik, C. P., and van Noordwijk, M. A. 1989. The special role of male *Cebus* monkeys in predation avoidance and its effect on group composition. *Behav. Ecol. Sociobiol.* 24: 265–276.

van Tuinen, M., Sibley, C. G., and Hedges, S. B. 2000. The early history of modern birds inferred from DNA sequences of nuclear and mitochondrial ribosomal genes. *Mol. Biol. Evol.* 17: 451–457.

Van Valkenburgh, B. 1999. Major patterns in the history of carnivorous mammals. *Ann. Rev. Earth Planet. Sci.* 27: 463–493.

Vanzolini, P. E., and Heyer, W. R. 1985. The American herpetofauna and the interchange. In Stehli, F. G., and Webb, S. D., eds., *The Great American Biotic Interchange.* New York: Plenum Press, pp. 475–487.

Vences, M., Glaw, F., Kosuch, J., Böhme, W., and Veith, M. 2001. Phylogeny of South American and Malagasy Boine snakes: molecular evidence for the validity of *Sanzinia* and *Acrantophis* and biogeographic implications. *Copeia* 4: 1151–1154.

Vergara-Aragon, P., Gonzalez, C. L. R., and Whishaw, I. Q. 2003. A novel skilled-reaching impairment in paw supination on the "good" side of the hemi-Parkinson rat improved with rehabilitation. *J. Neurosci.* 23: 579–586.

Vianna, D. M. L, and Brandão, M. L. 2003. Anatomical connections of the periaqueductal gray: specific neural substrates for different kinds of fear. *Braz. J. Med. Biol. Res.* 36: 557–566.

Vidal, N. 2002. Colubroid systematics: evidence for an early appearance of the venom apparatus followed by extensive evolutionary tinkering. *J. Toxicol.: Toxin Rev.* 21: 21–41.

Vidal, N., and David, P. 2004. New insights into the early history of snakes inferred from two nuclear genes. *Mol. Phylogen. Evol.* 31: 783–787.

Vidal, N., Delmas, A.-S., David, P., Cruaud, C., Couloux, A., and Hedges, S. B. 2007. The phylogeny and classification of caenophidian snakes inferred from seven nuclear protein-coding genes. *C. R. Biologies* 330: 182–187.

Vidal, N., and Hedges, S. B. 2002. Higher-level relationships of caenophidian snakes inferred from four nuclear and mitochondrial genes. *C. R. Biologies* 325: 987–995.

Vidal, N., and Hedges, S. B. 2004. Molecular evidence for a terrestrial origin of snakes. *Proc. R. Soc. Lond. B* 271: S226–S229.

Vidal, N., and Hedges, S. B. 2005. The phylogeny of squamate reptiles (lizards, snakes, and amphisbaenians) inferred from nine nuclear protein-coding genes. *C. R. Biologies* 328: 1000–1008.

Villeneuve, M., Kupers, R., Gjedde, A., Ptito, M., and Casanova, C. 2005. Pattern-motion selectivity in the human pulvinar. *NeuroImage* 28: 474–480.

Vuilleumier, P. 2002. Perceived gaze direction in faces and spatial attention: a study in patients with parietal damage and unilateral neglect. *Neuropsychologia* 40: 1013–1026.

Vuilleumier, P., Armoney, J. L., Driver, J., and Dolan, R. J. 2003. Distinct spatial frequency sensitivities for processing faces and emotional expressions. *Nature Neurosci.* 6: 624–631.

Waddell, P. J., and Shelley, S. 2003. Evaluating placental inter-ordinal phylogenies with novel sequences including RAG1, gamma-fibrinogen, ND6, and mt-tRNA, plus MCMC-driven nucleotide, amino acid, and codon models. *Mol. Phylogen. Evol.* 28: 197–224.

Walker, A. E. 1938. *The Primate Thalamus*. Chicago: University of Chicago Press.

Walker, D. L., and Davis, M. 2002. The role of amygdala glutamate receptors in fear learning, fear-potentiated startle, and extinction. *Pharmacol. Biochem. Behav.* 71: 379–392.

Walker, E. P., Warnick, F., Hamlet, S. E., Lange, K. I., Davis, M. A., Uible, H. E., Wright, P. F., and Paradiso, J. L. 1975. *Mammals of the World*, 3rd ed. Baltimore: Johns Hopkins University Press.

Ward, R., Calder, A. J., Parker, M., and Arend, I. 2007. Emotion recognition following human pulvinar damage. *Neuropsychology* 45: 1973–1978.

Ward, R., Danziger, S., and Bamford, S. 2005. Response to visual threat following damage to the pulvinar. *Curr. Biol.* 15: 571–573.

Watt, S. J., and Bradshaw, M. F. 2000. Binocular cues are important in controlling the grasp but not the reach in natural prehension movements. *Neuropsychologia* 38: 1473–1481.

Wayne, R. K., Benveniste, R. E., Janczewski, D. N., and O'Brien, S. J. 1989. Molecular and biochemical evolution of the Carnivora. In Gittleman, J. L., ed., *Carnivore Behavior, Ecology, and Evolution*. Ithaca, NY: Cornell University Press, pp. 465–494.

Weatherhead, P. J., and Blouin-Demers, G. 2004. Understanding avian nest predation: why ornithologists should study snakes. *J. Avian Biol.* 35: 185–190.

Webb, J. K., and Whiting, M. J. 2005. Why don't small snakes bask? Juvenile broad-headed snakes trade thermal benefits for safety. *Oikos* 110: 515–522.

Weiskrantz, L., Barbur, J. L., and Sahraie, A. 1995. Parameters affecting conscious versus unconscious visual discrimination with damage to the visual cortex (V1). *Proc. Natl. Acad. Sci.* 92: 6122–6126.

Weiskrantz, L., Warrington, E. K., Sanders, M. D., and Marshall, J. 1974. Visual capacity in the hemianopic field following a restricted occipital ablation. *Brain* 97: 709–728.

Weiss, R. A., and Wrangham, R. W. 1999. From *Pan* to pandemic. *Nature* 397: 385–386.

Werner, W. 1993. Nuerons in the primate superior colliculus are active before and during arm movements to visual targets. *Eur. J. Neurosci.* 5: 335–340.

Westby, G. W. M., Keay, K. A., Redgrave, P., Dean, P., and Bannister, M. 1990. Output pathways from the rat superior colliculus mediating approach and avoidance have different sensory properties. *Exp. Brain Res.* 81: 626–638.

Whishaw, I. Q. 2003. Did a change in sensory control of skilled movements stimulate the evolution of the primate frontal cortex? *Behav. Brain Res.* 146: 31–41.

Whishaw, I. Q., Sarna, J. R., and Pellis, S. M. 1998. Evidence for rodent-common and species-typical limb and digit use in eating, derived from a comparative analysis of ten rodent species. *Behav. Brain Res.* 96: 79–91.

Whishaw, I. Q., Suchowersky, O., Davis, L., Sarna, J., Metz, G. A., and Pellis, S. M. 2002. Impairment of pronation, supination, and body co-ordination in reach-to-grasp tasks in human Parkinson's disease (PD) reveals homology to deficits in animal models. *Behav. Brain Res.* 133: 165–176.

White, J. 2000. Bites and stings from venomous animals: a global overview. *Therapeutic Drug Monitor.* 22: 65–68.

Whiten, A., and Byrne, R. 1988. Tactical deception in primates. *Behav. Brain Sci.* 12: 233–273.

Wiener, S. G., and Levine, S. 1992. Behavioral and physiological responses of mother and infant squirrel monkeys to fearful stimuli. *Develop. Psychobiol.* 25: 127–136.

Wiens, J. J., Brandley, M. C., and Reeder, T. W. 2006. Why does a trait evolve multiple times within a clade? Repeated evolution of snakelike body form in squamate reptiles. *Evolution* 60: 123–141.

Wik, G., Fredrikson, M., Ericson, J., Eriksson, L., Stone-Elander, S., and Greitz, T. 1993. A functional cerebral response to frightening visual stimulation. *Psychiat. Res.: Neuroimaging* 50: 15–24.

Wildman, D. E., Uddin, M., Opazo, J. C., Liu, G., Kefort, V., Guindon, S., Gascuel, O., Grossman, L. I., Romero, R., and Goodman, M. 2007. Genomics, biogeography, and the diversification of placental mammals. *Proc. Natl. Acad. Sci.* 104: 14395–14400.

Wildman, D. E., Wu, W., Goodman, M., and Grossman, L. I. 2002. Episodic positive selection in ape cytochrome *c* oxidase subunit IV. *Mol. Biol. Evol.* 19: 1812–1815.

Wilkinson, M., Sheps, J. A., Oommen, O. V., and Cohen, B. L. 2002. Phylogenetic relationships of Indian caecilians (Amphibia: Gymnophiona) inferred from mitochondrial rRNA gene sequences. *Mol. Phylogen. Evol.* 23: 401–407.

Williams, C., Azzopardi, P., and Cowey, A. 1995. Nasal and temporal retinal ganglion cells projecting to the midbrain: implications for "blindsight." *Neuroscience* 65: 577–586.

Wilson, J. R., Hendrickson, A. E., Sherk, H., and Tigges, J. 1995. Sources of subcortical afferents to the macaque's dorsal lateral geniculate nucleus. *Anat. Rec.* 242: 566–574.

Wolfe, N. D., Switzer, W. M., Carr, J. K., Bhullar, V. B., Shanmugam, V., Tamoufe, U., Prosser, A. T., Torimiro, J. N., Wright, A., Mpoudi-Ngole, E., McCutchan, F. E., Birx, D. L., Folks, T. M., Burke, D. S., and Heneine, W. 2004. Naturally acquired simian retrovirus infections in central African hunters. *Lancet* 363: 932–937.

Wong-Riley, M. T. T. 1977. Connections between the pulvinar nucleus and the prestriate cortex in the squirrel monkey as revealed by peroxidase histochemistry and autoradiography. *Brain Res.* 134: 249–267.

Wong-Riley, M. T. T. 1994. Primate visual cortex: dynamic metabolic organization and plasticity revealed by cytochrome oxidase. In Peters, A., and Rockland, K. S., eds., *Cerebral Cortex, Vol. 10: Primary Visual Cortex in Primates.* New York: Plenum Press, pp. 141–200.

Wong-Riley, M. T. T., and Carroll, E. W. 1984. Quantitative light and electron microscopic analysis of cytochrome oxidase-rich zones in V II prestriate cortex of the squirrel monkey. *J. Comp. Neurol.* 222: 18–37.

Wong-Riley, M. T. T., Hevner, R. F., Cutlan, R., Earnest, M., Egan, R., Frost, J., and Nguyen, T. 1993. Cytochrome oxidase in the human visual cortex: distribution in the developing and the adult brain. *Vis. Neurosci.* 10: 41–58.

Wright, P. C. 1999. Lemur traits and Madagascar ecology: coping with an island environment. *Yrbk. Phys. Anthropol.* 42: 31–72.

Wu, W., Schmidt, T. R., Goodman, M., and Grossman, L. I. 2000. Molecular evolution of cytochrome *c* oxidase subunit I in primates: is there coevolution between mitochondrial and nuclear genomes? *Mol. Phylogen. Evol.* 17: 294–304.

Wüster, W., da Graca Salomão, M., Quijada-Mascarenas, J. A., Thorpe, R. S., and B.B.B.S.P. 2002. Origins and evolution of the South American pitviper fauna: evidence from mitorchondrial DNA sequence analysis. In Schuett, G. W., Hoggren, M., Douglas, M. E., and Greene, H. W., eds., *Biology of the Vipers*. Eagle Mountain, UT: Eagle Mountain Publishing, pp. 111–128.

Wüster, W., Ferguson, J. E., Quijada-Mascareñas, A., Pook, C. E., da Graça Salomão, M., and Thorpe, R. 2005. Tracing an invasion: landbridges, refugia, and the phylogeography of the Neotropical rattlesnake (Serpentes: Viperidae: *Crotalis durissus*). *Mol. Biol.* 14: 1095–1108.

Yoder, A. D., Burns, M. M., Zehr, S., Delefosse, T., Veron, G., Goodman, S. M., and Flynn, J. J. 2003. Single origin of Malagasy Carnivora from an African ancestor. *Nature* 421: 734–737.

Yoder, A. D., Cartmill, M., Ruvulo, M., Smith, K., and Vilgalys, R. 1996. Ancient single origin for Malagasy primates. *Proc. Natl. Acad. Sci.* 93: 5122–5126.

Yoder, A. D., and Yang, Z. 2004. Divergence dates for Malagasy lemurs estimated from multiple gene loci: geological and evolutionary context. *Mol. Ecol.* 13: 757–773.

Yokoyama, S., and Yokoyama, R. 1989. Molecular evolution of human visual pigment genes. *Mol. Biol. Evol.* 6: 186–197.

Yoshimura, N., Kawamura, M., Masaoka, Y., and Homma, I. 2005. The amygdala of patients with Parkinson's disease is silent in response to fearful facial expressions. *Neuroscience* 131: 523–534.

Yukie, M., and Iwai, E. 1981. Direct projection from the dorsal lateral geniculate nucleus to the prestriate cortex in macaque monkeys. *J. Comp. Neurol.* 201: 81–97.

Zaher, H., and Rieppel, O. 1999. The phylogenetic relationships of *Pachyrhachis probelmaticus,* and the evolution of limblessness in snakes (Lepidosauria, Squamata). *Earth Planet. Sci.* 329: 831–837.

Zamudio, K. R., and Greene, H. W. 1997. Phylogeography of the bushmaster (*Lachesis muta*: Viperidae): implications for neotropical biogeography, systematics, and conservation. *Biol. J. Linn. Soc.* 62: 421–44.

Zarbalian, G., Leung, P., Lower, L., Malkova, L., and Gale, K. 2003. Blockade of GABA-A receptors in the primate deep superior colliculus: behavioral effects Paper presented at the Society for Neuroscience 33rd annual meeting, New Orleans, Louisiana.

Zhao, B., Chen, H., and Li, B. 2005. Pattern motion and component motion sensitivity in cat superior colliculus. *NeuroReport* 16: 721–726.

Zimmerman, D. A., Turner, D. A., and Pearson, D. J. 1999. *Birds of Kenya and Northern Tanzania*. Princeton, NJ: Princeton University Press.

Zuberbühler, K., and Jenny, D. 2002. Leopard predation and primate evolution. *J. Hum. Evol.* 43: 873–886.

Acknowledgments

Many people have lent me their ear during my obsession with snakes and primates, and I would like them to know that their willingness to let me talk about my ideas helped me tremendously in formulating and fine tuning the theory. So, thanks very much to my birth family, Mom (Judith), Scott, Rand, Dann, Jeanne, and Leigh, and my late father, Earl; my family by marriage, Carol Young and John McDonald, and Truman and Anitta Young; my old friends, Hal Hinkle, Jan Kalina, and Randy Watson; my Fairfield friends, Berta Boegel (who told me that rubber snakes help keep cats out of the garden), Jim Rodgers, Solveig Seland, and Lois and David Suder; my Willett friends, Ted Lascher, Ellen Moratti, Liz Posner, Eileen Rendall (who sat next to me on the school bus and had no choice but to listen), Amy Slavich, and Spring Warren. Thanks also to Eileen, Spring, Sandy Harcourt, Nina Jablonski, Barbara King, Meredith Small, Kelly Stewart, and Louie Warren, who shared with this novice their insights about the process of writing and publishing an actual book.

This book was a multidisciplinary effort, and I relied heavily on many scientists to tell me when I was off the mark with details in their own disciplines. I hope that I have finally done a good job by them, but if there are errors in this book, it is not because of them. My great appreciation to neuroscientists John Allman, Vivien Casagrande, Leo Chalupa, Leah Krubitzer, Todd Preuss, Murray Sherman, Petroc Sumner, Daniel Ts'o, Deborah van der List, and Margaret Wong-Riley; psychologists Dick Coss and Don Owings; biological anthropologists John Fleagle, Karin Enstam Jaffe, Bill Kimbel, and Callum Ross; primate taxonomist Colin Groves; immunologist Nick Lerche; geneticist Derek Wildman; ornithologist Graham Martin; and multidisciplinarian Bob Martin. Fellow primatologists Rob Barton, Nate Dominy, and Sandy Harcourt also contributed in the same vein with their areas of special knowledge outside of primate behavioral ecology. Primatologists Dorothy Cheney, Janice Chism, Marina Cords, Donna Hart, and Charlie Janson all kindly shared unpublished information on primates' interactions with snakes.

Three scientists, in particular, helped in ways both unexpected and deeply appreciated: psychologist Arne Öhman, whose enthusiasm at a critical time encouraged me that there would be others who see what I see; neuroscientist Jon Kaas, who graciously agreed to the request of a stranger to critically read the penultimate draft; and herpetologist Harry Greene, who became a dear friend through shared stories, music, and old friends, and who also critically read the penultimate draft.

My past and present grad students, Dr. Rebecca Chancellor, Dr. Brianne Beisner, Dr. Noa Pinter-Wollman, John Bunce, Stephanie Etting, and Amy Porter, read the first draft I was willing to let anyone see and very diplomatically suggested alternative ways of presenting the material. I appreciate the time they took to help. Undergraduate students Susie Kang, Alissa McGowan, and Annie Laurie Norris also helped at crucial points, with Susie organizing and filing the papers that I read, Alissa editing, and Annie Laurie indexing. Alissa and Annie Laurie also read the entire first draft and gave me many constructive comments and suggestions.

The synthesis of the material and the process of writing demanded a certain intensity of thought and mental solitude. For the times I needed a respite, there was always Dave Alvin, Jackson Browne, Shawn Colvin, the late Dan Fogelberg, Nanci Griffith, Tim O'Brien, and Bonnie Raitt. They will probably never know how they helped me.

I am grateful to my editor, Michael Fisher, editor-in-chief at Harvard University Press, and the Board of Syndics at Harvard University Press for their support of this project. Their willingness to take on the book was, I'm sure, in no small part the result of generous reviews by Alison Jolly and an anonymous reviewer, both of whom also offered excellent insights that helped me streamline the book substantially. Thanks also to Michael for his editorial expertise in polishing the manuscript; to Anne Zarella, Michael's assistant, who helped with the details of publishing; and to Jennifer Seifert and John Donohue for their excellent copyediting skills. Kathy West created the illustrations that made the book more comprehensible, and it was a pleasure hashing out designs with her. The publication assistance fund, offered jointly by the Dean of the Division of Social Sciences in the College of Letters and Science at the University of California (UC) at Davis and the UC Davis Office of Research, helped defray the cost of the illustrations. I also thank the colleagues in my own Department of Anthropology for not questioning too loudly what snakes have to do with humans.

Finally, back to my own family: Truman Young deserves a huge acknowledgment for listening to me talk in brainspeak, for critiquing my "interminable" project, and for ignoring the stacks of journal articles, book chapters, and books that started like a mutated virus on the kitchen table and then quickly spread to all other flat surfaces in the house. Repeated cleanings could not get rid of the infection. And to Peter Young, my now teenaged son, thank you for taking it all in stride. And if you ever look back on your life and feel I didn't love you enough because I also loved my work, turn back to this page and soak it up: I love you more than a mama bear loves her cubs. I'm so glad you were around.

Index